Strategies to Cope with Risks of Uncertain Water Supply in Spate Irrigation Systems

Eiman Mohamed Fadul Bashir

Thesis committee

Promotor

Prof. Dr C.M.S. de Fraiture

Professor of Hydraulic Engineering for Land and Water Development

IHE Delft Institute for Water Education & Wageningen University & Research

Co-promotor

Dr I. Masih

Senior Lecturer in Water Resources Planning

IHE Delft Institute for Water Education

Other members

Prof. Dr R. Uijlenhoet, Wageningen University & Research

Prof. Dr N.C. van de Giesen, TUDelft

Prof. Dr C. Zevenbergen, IHE Delft Institute for Water Education & TUDelft

Dr A.M. Elkhidir Osman, University of Khartoum, Sudan

This research was conducted under the auspices of the SENSE Research School for Socio-Economic and Natural Sciences of the Environment

STRATEGIES TO COPE WITH RISKS OF UNCERTAIN WATER SUPPLY IN SPATE IRRIGATION SYSTEMS

Thesis
submitted in fulfilment of the requirements of
the Academic Board of Wageningen University and
the Academic Board of the IHE Delft Institute for Water Education
for the degree of doctor
to be defended in public
on Wednesday, 8 January 2020, at 3 p.m.
in Delft, the Netherlands

by

Eiman Mohamed Fadul Bashir
Born in Wad Medani, Sudan

Published by:
CRC Press/Balkema
Schipholweg 107C, 2316 XC, Leiden, the Netherlands
Pub.NL@taylorandfrancis.com
www.crcpress.com – www.taylorandfrancis.com

ISBN: 978-0-367-46582-7 (Taylor & Francis Group)
ISBN: 978-94-6395-157-9 (Wageningen University)
DOI: https://doi.org/10.18174/502551

To my mother, husband and daughters

To the memory of my beloved late father

Acknowledgments

I would like to express my sincere gratitude to my promoter Prof. Dr Charlotte de Fraiture for sharing her vast experience and knowledge with me and for providing continuous support and advice along the PhD years. Your conscientious support and insightful critical comments enormously helped in my professional growth and academic development. I am very honoured to have you as my promotor. Thank you for the guidance during the Ph.D. research.

I gratefully acknowledge with thanks, the precious supervision and guidance provided by my co-promoter Dr Ilyas Masih who accepted to join the supervision team in the third year. Many thanks for his knowledge, encouragement, patience and numerous time spent in keeping me in the right track. You were always present when I needed a short talk or loud thinking to guide me safely to concrete ideas.

I would also like to express my sincere gratitude to the Land and Water Development Core staff of the Water Science and Engineering Department, in particular Dr F.X. Suryadi for his great advices and experience sharing on irrigation models.

I would like to forward my sincere gratitude to the Dutch government and Netherland Fellowship Program (NUFFIC) for financing the expenses of this research. Further, I would like to acknowledge the Spate Irrigation Project funded by IFAD in Sudan for providing the platform to connect with the spate irrigation network in Sudan which facilitate research fieldwork.

Special thanks here to Gash Agricultural Scheme- Ministry of Agriculture, Gash River Training Unit & Hydraulic Research Centre- Ministry of Irrigation and Water Resources, Agricultural Research Corporation in Kassala, the Supreme Council of Water Users Associations in Gash, and the Gash Sustainable Livelihoods Regeneration Project – IFAD for their uncountable technical and logistical support. I would like to extend my deepest appreciation to Gash farmers and the authorities of Gash Agricultural Scheme; Mr. Kamal Ali and Mr. Mohamed Abdalla who provided all the facilities for data collection and database of the scheme. I am also thankful to Mr. Moawia Abdelfatah Mustafa from Kassal from Gash Research Station who provided a major part in data collection and laboratory analysis of soil samples.

I will take also this opportunity to forward my appreciation to Jolanda Boots, Niamh Mckenna, Anique Karsten, Loes Westerveen and other staffs of IHE Delft, for their kindly help and cooperation on addressing and managing all the administrative issues.

My sincerest gratefulness to all my friends and colleagues at IHE who are the real takeover treasure from the study experience at IHE. We shared the ups and down moments and the joy of having a paper being accepted and published.

Lastly, I would like to thank my husband Dr Raaed Mohamed Elhassan for his patience and complete support during the PhD process. Discussions based on his research background in the field of agriculture were extremely useful and appreciated. My beloved daughters Deena and Yasmin were extremely understanding, supportive, and tolerant during my home absence. Heartiest appreciation to my caring mother Khadiga and late father Mohamed Fadul who were always proud of me.

There are definitely some people I have missed to mention in this acknowledgement, but your contributions are greatly appreciated.

Eiman Fadul

Delft, December 2019

SUMMARY

Spate irrigation is a flood-based irrigation, a special type of irrigated agriculture that has been practiced in arid and semi-arid regions for centuries. The irrigation is based on diverting into the low lands the highly variable and unpredictable flash floods from valleys and ephemeral rivers using gravity force. Water supply in spate irrigation system is highly uncertain with likelihoods of receiving both extremely destructive flood and drought years. The uncertainty is inherent in the flooding time, volume of River flows, and in the annual irrigable area. Irrigation systems are frequently exposed to the impacts of climate variability and related extreme events such as floods and droughts which could result in large losses in agriculture productivity, assets and lives. Water supply risk and strategies to cope with climate variability in spate system needs to be addressed because spate irrigation contributes to the livelihoods and food security of marginalized populations in water scarce regions, where occasional floods are often one of the few sources of water for irrigated agriculture. Generally, studies on spate irrigation systems are limited. Particularly in Sudan, spate system have been neglected in national development plans and strategies. In addition, risk and coping strategies assessment in poor rural community systems, such as spate irrigation, has not been adequately addressed in the literature. In this context, this research aims to assess the main sources of risk and coping strategies due to uncertain water supply in spate-irrigated systems. The case study of this research is the Gash Agricultural Scheme (GAS) in eastern Sudan.

The research was conducted through the development of few methodological frameworks for risk and coping strategies assessment. Several methods were employed for data collection using field survey; questionnaires; and secondary data, and data analysis employing statistics, optimisation and modelling. For water supply risk assessment, a novel attempt is made to apply the SPRC (Source-Pathway-Receptors-Consequence) model originally developed for the flood risk management context, to a spate irrigation system in an arid region in Africa. The SPRC model, build upon the primary and secondary data, profoundly assisted in clearly comprehending and describing the sources of risks, propagation pathways, risk perceptions and consequences for the farmers, water user associations and water managers in the GAS. For coping strategies assessment, the Driving force-Pressure-State-Impact-Response (DPSIR) framework was used to identify strategies to cope with different water supply risks in the study area. Additionally, the mDSS4 (The MULINO Decision Support System) tool was employed to evaluate the effectiveness of the coping strategies. Then surface irrigation modelling using WinSRFR model was used to evaluate performance of locally developed practice for on-farm improvements for field design and water application. Performance of alternative designs and application times were simulated for different flood risks. Irrigation performance of different combinations was then examined using application efficiency, distribution uniformity, and adequacy criteria to obtain the best performing scenario. Last, a conceptual framework for establishment of real options in spate irrigation characterized by flexibility was developed with application in traditional, improved traditional and modernized spate irrigation systems.

The SPRC was a useful framework for analyzing risks at different spatial scales and for different stakeholders in the spate irrigation system. Based on limited knowledge and lack of flash flood forecasting systems, water flow to the irrigation system was unpredictable, uncertain with regard to volume, timing and duration. The stakeholders perceived flood risks as low flood, high flood, short flood, extended flood, early flood and late flood risks. Observations of flood events in the historical records of hydro-climatic data were categorized based on stakeholder's perceptions on threshold values for low flood, high flood, short flood, extended flood, early flood and late flood. Findings showed that farmers, WUAs and system managers perceived the risks from floods differently. The farmers were primarily concerned by low floods, while the WUAs were more disturbed by untimely floods. The system managers were most troubled by high and potentially destructive floods. The poor state of the infrastructure, lack of proper maintenance and suboptimal operation aggravated the consequences of unpredictable flows. Consequently, the resultant impacts were low crop yield, highly variable crop production and highly variable irrigated area.

The assessment of the effectiveness of existing coping strategies practiced by farmers, WUAS, and water managers revealed the most effective measures were crop management in terms of variety and change crop choices for farmers; pre-flood preparedness, risk sharing measures through water and land management during and after flood for WUAs; and flexibility in system operation by water managers. Unfortunately, the most effective measures were not the most adopted ones. The level of adoption is primarily related to the capacity of the farmers, WUAs and water managers to implement the measures without outside support.

Three strategies were investigated to evaluate performance of locally developed practice for on-farm improvements for field design and water application namely; time management strategy, improved field design with time management strategy and improved field design with flow management strategy. The second strategy resulted in the highest performance indicator values compared to other strategies. The adoption of improved field design with time management strategy resulted in the highest performance indicator values compared to other strategies, can save 40% of the current application time during large flood seasons, and 20% during medium flood seasons.

A conceptual, framework for flexibility consideration in spate irrigation was developed and applied. The framework consisted of four principle questions, eight main flexibility features and five sub-features that were found to adequately represent flexibility in spate irrigation systems. The conceptual framework demonstrated its beneficial use for the evaluation of spate irrigation system through its application on traditional, improved traditional, and modern spate systems to cope with high peak flood, low peak flood and untimely flood events.

A key contribution of this PhD thesis is the development of methodology frameworks for risk and coping strategies assessment in spate irrigation systems. This research developed approaches on how risks in spate irrigation systems could be assessed to enhance irrigation performance and equity to support farmers trapped in poverty and illiteracy. Additionally, water related risk management in low cost rural community system, such as spate irrigation system in arid and semi-arid zones, has been presented to the literature. Further, this research showed that spate irrigation performance could be optimized when proper set of coping strategies/real options are in place. Flexibility of spate irrigation systems are enhanced by adoption of real options to cope with variability and uncertainty of water supply.

TABLE OF CONTENTS

LIST OF FIGURES

LIST OF TABLES

1.1 SPATE IRRIGATION

Spate irrigation is a flood-based irrigation-a special type of irrigated agriculture that has been practiced in arid and semi-arid regions where evapotranspiration greatly exceeds rainfall (FAO AQUASTAT, 2010). The irrigation is based on diverting flash floods from valleys, rivers, riverbeds and gullies by gravity through irrigation canals to fields surrounded by earthen bunds (Lawrence and Van Steenbergen, 2005). Large volume of flood water induced by precipitation in the upper catchment is directed to low land and wadi areas in order to allow moisture storage in the soil profile to be utilized for crop production (Steenbergen and Haile, 2010, Haile *et al.*, 2006). Although unreliable water source, flash floods are often one of the few sources of water for irrigated agriculture in arid regions (Van Steenbergen, 1997, Asif and Islam-ul-Haque, 2014, Ghebremariam and Steenbergen, 2007).

Globally, the practice of spate irrigation is found in North Africa (Morocco, Algeria, Tunisia), East Africa (Sudan, Ethiopia, Eritrea, Somalia, Kenya, Tanzania), Middle East (Saudi Arabia, Yemen), West Asia & central Asia (Pakistan, Iran, Afghanistan), and in South America (Colombia, Bolivia, Ecuador, Peru) (Van Steenbergen, 1997, Finley, 2016, Zimmerer, 1995). The practice of flood irrigation itself is very old. In the Bolivian Andes in Latin America, Zimmerer (1995) reported that an intensive spate irrigation system was in use at about AD 719 and that it was in operation as early as 3500 years before present (BP). In Pakistan and Yemen, the history of spate irrigation dates back to over 5000 years, and Pakistan has the largest spate irrigated area in the world (Steenbergen *et al.*, 2010). Traditional spate irrigation systems have been practiced in Kenya since 400 years ago at the lower reaches of Tana River. Currently, spate irrigated area covers 3.0 million hectares of irrigated land around the world (Steenbergen *et al.*, 2011) to provide multiple uses such as crop production, horticulture, groundwater recharge, rangeland development, forestry and small-scale water storage for domestic and livestock water supply. Spate irrigation provides the source of livelihood and food security for about 9-13 million people in the world (Steenbergen *et al.*, 2010). Although spate irrigation is one of the oldest irrigation systems, local practices that were developed along the history with different unique experiences, are (still) less documented and not well disseminated. Figure 1.1 shows approximate spate irrigation areas in some of the countries as reported by different authors. Since the potential area is larger, the maximum reported areas were only shown.

In the semi-arid regions of Sudan, farmers divert flash floods from intermittent seasonal rivers such as Gash, Toker and Khor Abu Habil to sustain agricultural production in eastern and the western part of the country using spate irrigation technology. It was first developed in 1872 for cotton production in Toker spate system in far eastern Sudan and later developed in 1924 in the Gash agricultural scheme (GAS) in east Sudan.

Steenbergen *et al.* (2010) reported an area under spate irrigation amounts to 132,000 ha in Sudan, however, the potential is much larger.

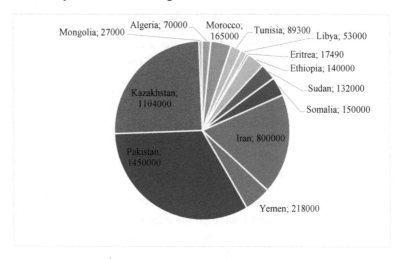

Figure 1.1 Approximated spate irrigation areas worldwide (ha).

1.2 RISK OF UNCERTAIN WATER SUPPLY

The occurrence of flash flooding is the most top-ranked events among natural disasters in terms of both the number of people affected globally and the proportion of individual fatalities (Marchi *et al.*, 2010, Jonkman, 2005). However in several areas in the world those destructive flash floods are managed and diverted from normally dry riverbeds and spread gently over agricultural land for crop and livestock production (Steenbergen et al., 2011, Haile et al., 2011). This operation is termed spate- irrigated agriculture.

Variability in weather elements is the principal source of fluctuations in global food production, particularly in the semi-arid tropical countries of the developing world (Aggarwal *et al.*, 2010). Climate change and variability affected the crop production of some staple crops and future climate change threatens to exacerbate this (FAO, 2018). Exposure to climate variability and extremes, poses substantial risks to people living in the Sudan(Elasha *et al.*, 2005),with an expected threat to exacerbate poverty and create new poverty pocket (Sabine, 2014). Exposure to risk prevents farmers from safely planning ahead and making investments (Binswanger and Sillers, 1983). Consequently, external parties are reluctant to invest in agriculture because of the uncertainty about the expected returns (Steenbergen *et al.*, 2011).

Sudan is one of the most fragile and vulnerable areas to climate change and climate variability in Africa (Mohamed *et al.*, 2016, USAID, 2016). It extends in the Sahel region which has a significantly warmer weather with a systematic decrease in rainfall (Lucio *et al.*, 2012, Brooks, 2004). The Sudanese economy is predominantly agricultural based, which contributed together by an average 44.4% in the GDP. More than 70% of population is directly dependent on climate-sensitive resources for their livelihood (HCENR, 2003). The country is facing sever water scarcity, which affects significantly the country's development (Aimar, 2017) and in particular the agricultural sector.

Recently, GIEWS (2018) reported 6.2 million people vulnerable to sever localized food insecurity due to conflict and weather shocks. Food insecurity in Sudan is directly linked to climatic and non-climatic factors among which climate change & variability (Osman-Elasha *et al.*, 2006), conflicts & internal displacement population (Gundersen, 2016), uncertainty in agricultural production (Muli *et al.*, 2018) and low crop productivity (Siddig and Babik, 2017).

Water supply in spate irrigation systems is characterized by very high natural hydrologic variability, heavy sediment load, high uncertainty of floods, the need to capture short-duration floods, the special operation and maintenance approaches needed, and the exceptional nature of the water rights (Steenbergen and Haile, 2010). There is considerable uncertainty in the timing and the volume of floods (Fadul et al., 2017; Van Steenbergen, 1997) which resulted in variability in annual irrigated area, low crop productivity and poor operation and maintenance.

1.3 STRATEGIES TO COPE WITH RISK

Water management in spate-irrigated schemes is a complex and multifaceted process. Therefore, farmers and other actors have developed local knowledge and practices over the years to manage and use the un-predictable water supply to produce crops such as cotton, sorghum, castor, sunflower and vegetables. The development of spate irrigation systems are mostly based on traditional knowledge and experience gained along the years. Issues such as land and water distribution, size and angle of diversion structure are still evolving in different places according to local circumstances. In addition, there is a lack of context specific guidelines for the diverse spate irrigation systems, though only general broad guidelines have been developed by FAO organization in 2010 (Steenbergen and Haile, 2010). Hence, there is still a lack of evidence-based and context specific knowledge on complex spate systems (Erkossa, 2014).

1.4 PROBLEM STATEMENT

Similar to other irrigated areas in Sudan, crop yields from spate-irrigated area are low and constrained by exposure to climate variability and extremes that poses substantial risks to people, and their properties (Elasha *et al.*, 2005). Exposure to water supply risks prevents farmers from safely planning ahead and making investments (Binswanger and Sillers,

1983). Consequently, external parties are reluctant to invest in agriculture because of the uncertainty about the expected returns (Steenbergen *et al.*, 2011). The spate irrigation has been less recognized in the literature of irrigation technologies and only few authors have discussed local cases focusing on system descriptions and recommendations for future development without detailed investigation of sources of risk and existing strategies to cope with them. This might be attributed to the focus of scientific research and approaches on flood risk management to urban system targeting protection of cities, towns and residential areas with high economic value. Therefore, there is a lack of research on flood risks in low cost rural community system such as spate irrigation system in arid and semi-arid zones. More research is needed to explore on: how farmers and their institutions perceive water supply risks and measures to deal with it; the major consequences of uncertain water supply on the system; the different strategies adopted to cope with uncertain water supply; the linkage between different water supply risks and the strategies developed; the effectiveness and adoption rates of existing strategies; and the performance of an existing and alternative strategies.

1.5 OBJECTIVES

The main objective of this research is to assess the risks and coping strategies to cope with uncertain water supply in spate irrigation to contribute towards achieving sustainable livelihood farming communities, taking the Gash agricultural scheme (GAS) in Sudan as a case study.

The specific research objectives of this study are as follows:

1. To assess the main elements of risks due to uncertain water supply that have significant impacts on spate irrigation performance in the GAS.

2. To evaluate the effectiveness of coping strategies and practices that have been developed over years to cope with uncertain water supply in GAS.

3. To identify alternative locally feasible measures that would address the different level of hydrological events and cope with variability of water supply and enhance irrigation performance.

4. To establish a conceptual framework for adoption of real option that enhance system flexibility to cope with variability and uncertainty of water supply.

1.6 CASE STUDY AREA

The Gash Agricultural Scheme (GAS) was selected as the main study area because it captures important hydro-climatic and management-related characteristics of a typical spate irrigation system, and provides an important value for the local economy and livelihood of poor communities in east Sudan (Ngirazie *et al.*, 2015).

The GAS is the largest spate-irrigated area located in east Sudan (15˚27ʹN, 36˚24ʹE). The ephemeral Gash River is the main water source for the GAS. It originates from the Eritrean and Ethiopian highlands where it is called the Mareb River. Flash floods from Gash River, occurred from mid-June to end September, are the source of irrigation for the GAS. The river is characterized by large annual flow variability, high sediment concentration and responds rapidly to storm rainfall in the upper catchment. The peak flows have been estimated up to 1000 m^3/s at Kassala Bridge. The maximum and minimum annual flow, which occurred in the years 1983 and 1921, were recorded as 1430×10^6 and 140×106 m^3/year, respectively. Sediment concentration in the Gash River may exceed 60,000 ppm during high flood (Zenebe *et al.*, 2015b).

The climate in the study area is arid to semi-arid with an average temperature of 31˚C during May-August and 22˚C during September-January. Rainfall is highly seasonal occurring between July and October. The average annual rainfall ranges from 260 mm in the southeast to less than 100 mm in the northwest of the GAS (IFAD, 2003). High evapotranspiration rates of up to 2,000 mm/year diminishes the effectiveness of rainfall as the main source of water supply for crop production, and makes irrigation a necessary source to compensate for the evapotranspiration deficit (FAO, 2016).

The GAS system is designed to irrigate a total area of 100,000 ha using the spate floods from the Gash River. The average annual irrigated area is approximately 30,000 ha which is greatly depends on the number of flood events, their duration and the farmers' capacity to timely divert, operate and manage the flood water. In the past, the GAS system has been successfully operated for cotton crop production at the upper and middle parts of the field, while occasionally the lower end of the field has been used to produce sorghum for home consumption.

Since 1980, irrigated agriculture in GAS has been on the decline due to drought periods, changes in river course and changes in the institutional set-up. Additionally, the majority of the farming communities living in this marginal area were nomadic but forced to settle as a result of droughts, war and decline in the vegetation cover. This has resulted in large demands on food supply and pressure on the limited water sources and infrastructure to include displaced population into farming practice. Therefore, farming communities have had to change their source of living to subsistence farming and livestock grazing.

Water user association (WUAs), established in 2004, further institutionalized the basis for land and water distribution in the GAS (Abdelgalil and Bushara, 2018).The cropping pattern has changed from cotton to sorghum as the main crop produced in GAS. Sorghum, a deep-rooted and drought-resistant crop, is currently grown in most of the area. There are approximately 45,000 tenant farmers organized in 92 WUAs (on average 433 farmers per WUA). There are three main stakeholders responsible for irrigation water management: 1- water managers who are responsible for the operation and maintenance (O&M) of the main canals and diversion structures, 2- WUAs who are responsible for the O&M of secondary canals and offtakes, and 3- farmers who manage the field canals and on-farm water distribution structures such as spurs and field embankments.

The irrigation system consists of seven irrigation intakes to irrigate six irrigation blocks through a network of main canals, secondary canals and field canals. Secondary canals divert water from the main canals to fields through embankment breaching or via field canals. On average, each irrigation block is divided into 35 ranges from 250 to 1,250 ha, and is irrigated by one secondary canal to serve a group of farmers (300–600 farmers). Kassala Block at the upstream and Metateib Block at the downstream end of GAS were selected for investigation (Figure 1.2).

The command area is irrigated based on a two-year land rotation so that 50% of the area is irrigated in one season while the remaining area are left fallow to be irrigated next season. Similarly, within the flood season, the irrigation fields are scheduled in: 1st irrigation fields, which are irrigated continuously for a period of 25–30 days (10th July–10th August); and 2nd irrigation fields which are irrigated after irrigation stops from 1st fields (10th August–10th September). In year 2 the other fallow irrigable area is targeted to irrigate new sets of 1st and 2nd irrigation fields.

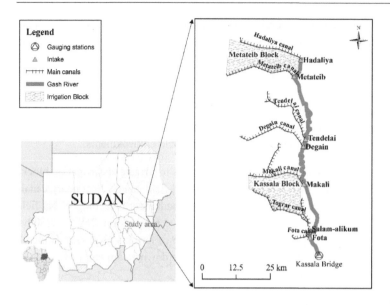

Figure 1.2: Gash agricultural scheme in Sudan (The study area).

1.7 METHODS

First, the research conducted field surveys and interviews with farmers, water user associations (WUA), water managers and relevant institutions and policy makers in the Gash spate irrigation system in Eastern Sudan (GAS) to identify risks faced by local communities and their institutes in managing irrigation water under variable and uncertain water supply. Then, a novel approach was developed to analyse risks by applying the SPRC (Source-Pathway-Receptors-Consequence) model, originally developed for flood risk management in coastal zones, to a spate irrigation system in semi-arid region in Africa. The SPRC model, by building on primary and secondary data, assisted in a profound comprehension and description of the sources of risks, propagation pathways, risk perceptions and consequences for the farmers, water user associations and water managers in the GAS. Second, field surveys and interviews with farmers, water user associations (WUA), and water managers was conducted in GAS to identify local strategies developed. The effectiveness of coping/adaptation strategies was evaluated using the mDSS4 (The MULINO Decision Support System) tool which was based on the Driving force-Pressure-State-Impact-Response (DPSIR) framework. Third, focusing on field design and water management practices, local and alternative measures for on-farm improvements were investigated in GAS using surface irrigation modelling and field measurements. The research applied the hydraulic simulation model WinSRFR 4.3.1 to examine the performance using application efficiency, distribution uniformity, and adequacy criteria. Finally, this research developed a conceptual framework for establishment of flexible real options in spate irrigation. A set of flexibility features had been recommended for evaluation of different types of spate irrigation systems. Figure 1.3 represents the methodological framework of the PhD research.

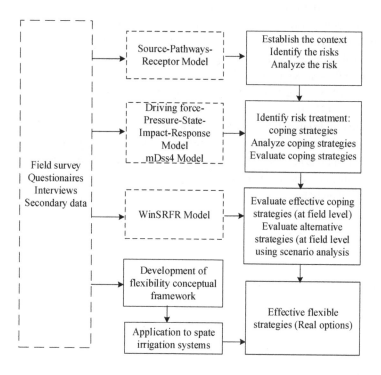

Figure 1.3: The methodological framework of the research.

1.8 THESIS OUTLINE

This thesis was structured in six chapters including the introduction and conclusions. Two peer reviewed publications and two submitted papers were developed during the study and were presented as individual four chapters. The research papers, following the four research objectives, contributed to the limited knowledge and understanding of the spate irrigation systems. In chapter 1, a general introduction, problem description, objectives, and overall research methodologies were presented. Chapter 2 presented an assessment of the most important risks due to uncertain water supply in GAS. It explored the risks of using unpredictable flash floods for irrigation. The research also identified how risks were perceived by individual farmers at field level, water user associations at secondary canal systems and water managers at primary systems. Moreover, a novel attempt was made to apply the Source-Pathway-Receptor-Consequence (SPRC). Chapter 3 evaluated the effectiveness of adaptation/coping strategies developed by farmers, WUA's and GAS managers to cope with high, low and untimely flood events in GAS scheme. The methods employed the mDSS4 tool which was based on the Driving force-Pressure-State-Impact-Response (DPSIR) framework to evaluate the effectiveness of strategies. Using data from interviews with 101 randomly selected farmers, 17 water user associations (WUAs), and 7 system water managers, the effectiveness with the rate of adoption were addressed. Chapter 4 focused on field level adaption strategies to develop alternative strategies for high, medium and low floods. We examined the performance of improved field design strategies to manage variable irrigation water supply and application time in the GAS where open-end border irrigation was practiced to irrigate large fields. Chapter 5 compared the flexibility of different types of spate irrigation systems in coping with water supply variability and uncertainty through adoption of real options. A conceptual framework was for real options in spate irrigation was introduced in this research. Chapter 6 presented the conclusions, and recommendations for further research. Appendices present supplementary materials.

2

RISKS OF UNCERTAIN WATER SUPPLY IN SPATE IRRIGATION[1]

This chapter explores the sources of risks, propagation pathways, risk perceptions and consequences for the farmers, water users' associations (WUAs) and water managers in the Gash Agricultural Scheme in Sudan using the Source–Pathways–Receptor and Consequence (SPRC) framework. Farmers in flood-based irrigation systems face great uncertainties with respect to water supply. The main source of risk is the extreme variability of rainfall, causing unpredictable flows regarding volume, timing and duration. The farmers, WUAs and system managers perceive the risks from floods differently. The farmers are primarily concerned by low floods, while the WUAs are more disturbed by untimely floods. The system managers are most troubled by high and potentially destructive floods. The poor state of the infrastructure, lack of proper maintenance and suboptimal operation aggravate the consequences of unpredictable flows. Consequently, the result is low and highly variable crop production. Besides paying attention to infrastructure improvement and regular operation and maintenance activities, more efforts in institutional arrangements and policy support could play an important role in coping with the risks indicated. The SPRC appeared to be a useful framework for analysing risks at different spatial scales and for different stakeholders in the spate irrigation system studied.

[1] This chapter has been published in: FADUL, E., FRAITURE, C. D. & MASIH, I. 2018. Risk Propagation in Spate Irrigation Systems: A Case Study from Sudan. *Irrigation and Drainage,* 67, 363-373.

2.1 INTRODUCTION

Flash floods in semi-arid rural areas are often associated with risks to human and animal life, and damage to or destruction of infrastructure and property. On the other hand, flood based irrigation also offer opportunities to the poor farming communities (Steenbergen and Haile, 2010). Flash floods can contribute to the livelihoods and food security of marginalized populations in water scarce regions, where occasional floods are often one of the few sources of water for irrigated agriculture (Asif and Islam-ul-Haque, 2014, Ghebremariam and Steenbergen, 2007).

Spate irrigation is one type of flood based farming that makes use of highly variable and seasonal flash floods in ephemeral rivers (also referred to as 'spate'). Flash floods sometimes come in very large quantities and do huge damages. At other times, these floods may reduce to small and unusable flows for irrigation.

Due to their erratic and variable nature (both in space and time), flash floods are difficult to monitor with conventional discharge measurement systems (Creutin and Borga, 2003, Borga *et al.*, 2008, Borga *et al.*, 2011). Similarly, modelling rainfall-runoff processes is also challenging in these systems, due to lack of data and the high degree of complexity of hydrological process (Rozalis *et al.*, 2010). These features pose additional challenges in the effectiveness of early warning systems and real time peak flow measurements (Perks *et al.*, 2016).

On the risk management, research indicates that understanding farmers' risk perceptions could significantly contribute towards formulating and implementing appropriate adaptation measures and policies (Botzen *et al.*, 2009, Adelekan and Asiyanbi, 2016). In addition, the lack of knowledge in this regard may lead to weak social acceptance of a suggested strategy (Touili *et al.*, 2014). Moreover, several factors potentially affect the perception of climate variability: personal experience, political interests and institutional support (Broomell *et al.*, 2015, Spence *et al.*, 2011, Niles and Mueller, 2016). For example, in a study in New Zealand, NilesMueller (2016) demonstrated that perceptions may be influenced by both personal beliefs and interpretations of climate change occurrences. Risk perceptions are personal, not static and may change over time, hence adaptation strategies should also be flexible and continuously updated (Duinen *et al.*, 2015).

Therefore, a thorough understanding of stakeholders' risk perceptions is inevitable if suitable policy actions are to be formulated. For example, farmers' perceptions on water supply risks contribute to shaping their farming practices and associated risk management strategies. Understanding these facts could help decision makers optimize limited resources, focusing on the real risks faced by local stakeholders (e.g. farmers) and their institutions. Few studies conducted in Africa investigated the farmers' general perceptions on risks related to climate change (Deressa *et al.*, 2011, Fosu-Mensah *et al.*, 2012, Maddison, 2007) or more specifically, in terms of increase or decrease in the climatic parameters (Duinen *et al.*, 2015). In a study in the rural Sahel in Africa, it was pointed that despite individual farmers being aware of climate variability, farmers' response in group interviews were influenced by strong narratives on climate (Mertz *et al.*, 2009).

Moreover, there is a limited knowledge about water supply risks faced at the farm level for many irrigation systems across the world (Nicholas and Durham, 2012); a similar situation could be found in the spate irrigation systems in Sudan . Thus, more research is needed to explore how farmers and their institutions perceive climate-related risks in flood-based irrigated agriculture; what are the major consequences of uncertain water supply for these systems and the dependent stakeholders? This paper addresses these knowledge gaps by exploring the risks of using unpredictable flash floods in irrigation, and farmers' and managers' perceptions in the spate irrigation based Gash Agricultural Scheme (hereafter referred to as GAS) in Sudan. The research identifies how risks are perceived by individual farmers at field level, WUAs in secondary canal systems and water managers in irrigation main canal systems. Moreover, a novel attempt is made to apply the SPRC model (discussed below), originally developed for the flood risk management context, to a spate irrigation system in an arid region in Africa. The SPRC model, built upon primary and secondary data, profoundly assisted in clearly comprehending and describing the sources of risks, propagation pathways, risk perceptions and consequences for the farmers, WUAs and water managers in the GAS. The empirical knowledge generated in this research could potentially contribute to the development of suitable policy actions to transform highly variable, unpredictable and underperforming spate irrigation systems into more resilient and productive ones, which are better able to cope with the risks posed by the high degree of climatic and hydrological variability.

2.2 DATA AND METHODS

2.2.1 The SPRC conceptual model

The SPRC model is a conceptually method to describe systems and processes (Horrillo-Caraballo *et al.*, 2013) through the representation of a particular source of risk, its propagation and consequences (Narayan *et al.*, 2011, FLOODsite, 2009, Narayan *et al.*, 2012). In coastal risks, it serves as a powerful tool for envisioning and contextualizing structural mitigation options (Touili *et al.*, 2014). This model explores the pathway between hazard —or source of risk—and receptors. The pathway is the physical structure by which the receptor is linked to the source. The existence of a pathway, linking source to the receptor, is a condition for risks to occur. Moreover, the use of system diagrams allow for a comprehensive description of the state of the system under investigation, its elements and their (spatial) linkages (Narayan *et al.*, 2012). The approach is further advanced by combining the system diagram and SPR model for the description of flood plain systems in the coastal areas in Europe (Narayan *et al.*, 2014). Recently, Narayan *et al.* (2015) included Bayesian Network approach in the SPRC application, which helped to identify critical flood system components and quantify inundation probabilities.

Although past development of the SPR model and its combination with other approaches were successful in characterization of coastal floodplain system and provide rapid risk assessment, they did not include the different risk perceptions and risk consequences for different stakeholders at various spatial levels. Further, neither of the past studies have focused on risks related to uncertain water supply in flood-based irrigation schemes. In this study, we adapted and applied the SPRC model to a spate irrigation system (the GAS), and have included new dimensions that act as the pathways to risk propagation. Pathways include stakeholders' perceptions of risks, flood variability in terms of volume, duration and timing, infrastructure and soft measures (operation and maintenance and institutional arrangements).

2.2.2 Sampling method

There are approximately 65,000 registered farmers distributed in six irrigation blocks in the GAS. In total, we interviewed 101 farmers, adopting a stratified sampling technique from two irrigation blocks: the Kassala Block at the upstream and Metateib Block at the downstream end of GAS. The interviewed farmers were distributed in head, middle and tail locations along the selected irrigation fields. The selected fields were located upstream, midstream and downstream relative to the main canal at each irrigation block (Figure 2.1). Further, key informants were interviewed from 17 WUAs functioning in the GAS. The water managers, staff from government agencies and local administration were also consulted. The fieldwork took place between June–August 2014 and January–February 2015.

In addition, the hydro-climate variability was assessed using secondary data from historical records of river flows and precipitation. There are six gauging stations for measuring discharge. The head–discharge rating curves are used for estimating flows from the observed water levels. The data of the most accurate and reliable station, the Kassala Bridge station, was used for the analysis. Furthermore, the relevant information on agricultural production and salient features of irrigation and farming systems was collected from the corresponding government agencies.

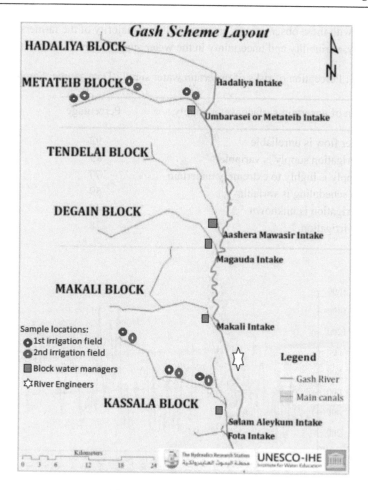

Figure 2.1: Sample locations.

2.3 RESULTS AND DISCUSSION

2.3.1 Sources of risks based on the hydro-climatic data

The main source of risk in the GAS is the variability and uncertainty of water supply. The erratic nature of rainfall occurring in the upper catchments in Eritrea and Ethiopia is transformed into unpredictable river discharges. Furthermore, river flows show large inter-annual variability (Figure 2.2). The minimum flow of 140×10^6 m^3/year was recorded in 1921, whereas the maximum flow of 1430×10^6 m^3/year occurred in 1983. Figure 2.3 illustrates long-term averages of rainfall and River flows.

Consistent with these observations, an overwhelming majority of the farmers considered unreliability, variability and uncertainty in the water supply as the main risk (Table 2-1).

Table 2-1: Perception of risks of uncertain water supply-Data source: Farmer survey

Perception of uncertain surface water supply	Percentage
Gash River flow is unreliable	92
Annual irrigation supply is variable	89
Water supply is highly to extremely uncertain	77
Irrigation scheduling is variable	59
Start of irrigation is unknown	43
Untimely irrigation	28

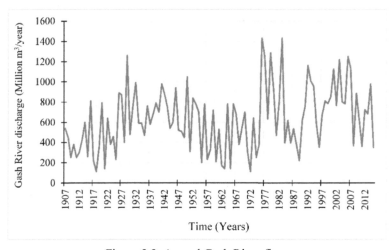

Figure 2.2: Annual Gash River flow.

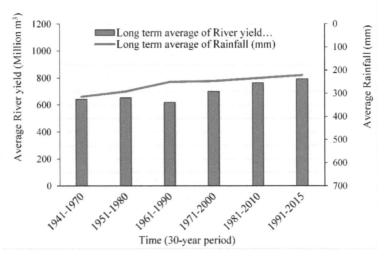

Figure 2.3: Variability of water supply in the study area.

2.3.2 Sources of risk based on farmers' perceptions

As pointed out earlier, the highly variable and unreliable water supply is the main source of risk, as stated by the farmers and WUAs members. Specifically, these risks are manifested as: low and high floods, late and early floods, short and long floods. Following on from these observations, the respondents were asked to suggest thresholds to classify these floods. These inputs helped with categorization of these events (Table 2-2 and Figures 2.4 (a–c).

Table 2-2: Classification of floods according to farmers. Data source: Field survey

Quantity (Mm^3/year)	Timing	Duration (days)	Water level (m)
Large: >1000	Early start: before 25 June	Short: < 65	Destructive: > 507
Good: 800-1000	Late start: after 15 July	Long: > 90	Average: 505.5
Average: 400-800	Early end: before 7 Sept		
Low: <400	Late end: after 7 Oct		Low: < 504

A few important observations could be made after applying the suggested classification thresholds to the historic river flow records. First, the number of long season's floods exceeded the short-season ones (Figure 2.4a). Second, early floods were more frequent than late ones (Figure 2.4b). Third, high floods outnumbered low flood events (Figure 2.4c).

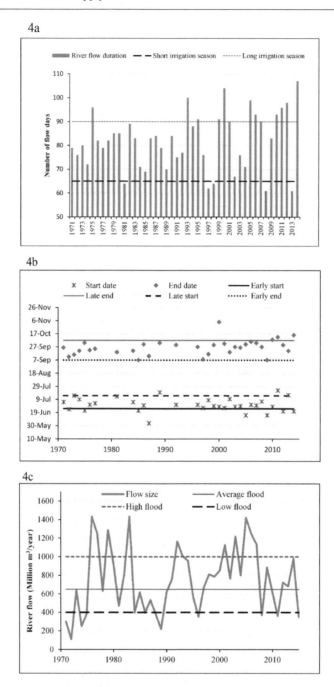

Figure 2.4 a,b,c: Comparison of historical River flood thresholds- Data source: GRTU and interviews.

Moreover, the farmers were asked to assess the likelihood of high/low, late/early and long/short floods. The occurrence of high floods was considered unlikely by three-quarters of the respondents (Figure 2.5). This assessment was in contradiction to the observed flow data (Figure 2.4a), which meant that the farmers had a tendency to underestimate the occurrence of high floods. On the other hand, the occurrence of low floods seems to be overestimated. Some 78% of farmers perceived low floods to occur more than 50% of the time. Similarly, the occurrence of late floods was overestimated, as half of the farmers believed that they occur more than 50% of the time. The survey revealed that the experience of farmers with water supply failures for their individual fields had strongly influenced their perceptions about the risks posed by different types of flood events. This affirmed the findings of earlier studies (Niles and Mueller, 2016), suggesting the strong influence of personal experience on risk perceptions.

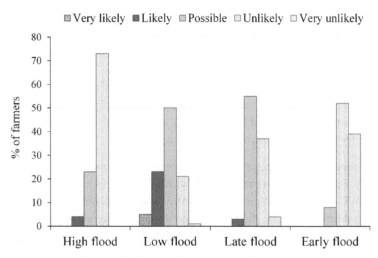

Figure 2.5: Farmers' perception of risk sources.

2.3.3 Pathways linking risks to receptors

Pathways are the linkages between source (i.e. uncertain water supply) and receptors (i.e. stakeholders who bear the risks and their consequences, such as farmers, WUAs and scheme managers). Pathways transfer and propagate risks to the receptors and constitute the physical infrastructure such as intakes, control structures and canals through which the uncertain water supplies reach the farmers' fields. Pathways also include the institutions and processes through which the physical infrastructure is operated and maintained.

Physical infrastructure. The physical infrastructure consists of the **headworks** located on the left bank of the Gash River for diversion of irrigation water to the scheme. The **primary system** includes the distribution system of main canals and structures to convey the water supply from the headworks to the secondary system. The **secondary system** includes secondary canals and offtake structures to deliver irrigation water to irrigation fields and field canals. The **on-farm system** conveys water to farm units and aims to ensure uniform distribution within irrigation fields through lateral embankments, field canals and spurs.

Lack of proper infrastructure for water capture, conveyance and distribution exacerbates the effects of an uncertain water supply and gives rise to water conflicts and inequity. For example, large command areas in the downstream Metateib block lack proper irrigation infrastructure. Consequently, farmers are forced to violate land rotation rules by cropping on the same fields for several years. This results in loss of soil fertility and low productivity. In the upstream Kassala block, the suboptimal location and lack of proper river training works (at the Fota intake) lead to excessive sedimentation along the diversion canals, which results in poor water distribution (Zenebe *et al.*, 2015a).

Operation and maintenance processes. The operation and maintenance (O&M) system includes water allocation management based on agreed water-sharing rules, water services such as canal operation, field offtake operation, water monitoring, and maintenance by undertaking preventive, reactive and daily activities needed to restore system performance. Because of the highly variable water supply and heavy sediment and debris load, adequate O&M of the irrigation system is a challenging and crucial undertaking for timely and equitable water delivery.

The lack of proper O&M propagates and exacerbates the adverse effects of an uncertain water supply. For example, maintaining field structures requires heavy equipment that is beyond the WUA's capacity (Ngirazie *et al.*, 2015) and can only be provided in a limited number by scheme managers. Political and tribal interference give rise to inequitable access to equipment. Because of limited and late budget allocations, the scheme follows an emergency approach, which considers only very critical and limited locations for yearly maintenance before the flood season. This approach results in the recurrence of breaching and poor conveyance capacity of irrigation networks, thus aggravating the effects of a highly variable water supply.

In the absence of an early flood warning system, operation of this spate irrigation system is a major challenge. In addition, there is a lack of reliable flow-measuring stations at the GAS intakes, which further complicates well-informed operational decisions. Thus, in practice, operation of the intakes and control structures downstream of the intakes relies heavily on the experience and sound judgement of gate operators. Errors incurred in this process may lead to negative consequences such as infrastructure damage, poor irrigation or irrigation failure.

2.3.4 Receptors – Those who bear the risks

The receptors of water-related risks in the GAS are found at three levels in the irrigation system: (i) water managers, who are responsible for the O&M of the primary system, (ii) the WUAs who are responsible for the O&M of secondary systems, and (iii) farmers who manage the field canals and on-farm water distribution through spurs and field embankments. These receptors at different levels are concerned with different types of floods according to their role and interest.

Farmers. Most of the farmers were mainly concerned with low floods because these events lead to insufficient water for their fields and poor water distribution along the main canal and within the irrigation fields. This point was supported by 90% of the farmers, who claimed that it is 'likely' or 'very likely' that poor water distribution occurs, as opposed to only less than 20% who thought that 'over-irrigation' is likely. These responses were also mirrored by farmers' perceptions (stated above) on the higher probability of low rather than high floods.

Poor water distribution is caused by very large field sizes and limited field infrastructure. The size of irrigation fields varies from 250 to 1,250 ha in the GAS, with 200–900 farmers sharing the same inlet. Poor irrigation occurs when (part of the) fields receive flood water in volumes less than the crop water requirement or when less than 60% of the planned area is not irrigated. Irrigation failure occurs when irrigation water fails to be diverted to the prepared fields.

Low floods passing through unlevelled fields lead to poor water application uniformity, which leads to the formation of dry 'islands' in the higher parts and waterlogging in the lower parts of fields. The poor management of irrigation water resulted in 80% of farmers claiming that they had failed to get irrigation water on their farms at least once in the past 5 years.

On the other hand, due to the absence or weak nature of field infrastructure to control the flows to fields and proper drainage, high and long-duration floods often lead to breaching, over-irrigation and damage to standing crops. On the other hand, the early arrival of floods may lead to losses of water because the fields were not prepared.

Differences between the upstream Kassala and downstream Metateib blocks. The perceived risks differ substantially between the farmers from the upstream Kassala block and the downstream Metateib block, as well as within the same block due to their location. Two-thirds of the farmers in the downstream fields at Kassala have experienced irrigation failure in the past, compared to half of the farmers in the upstream fields. All the farmers at Metateib experienced irrigation failures from time to time, particularly in the middle fields. Unfortunately, very few farmers could cultivate their lands in the downstream part of Metateib because of the incomplete or lacking canal and poor water conveyance. Hence, several fields were left abandoned, as the water did not reach there, as opposed to the tail-end farms in the upstream Kassala block, where some fields are even exposed to over-irrigation and accumulation of sand and debris.

Furthermore, farm intakes and canals located at the head of the Kassala block were mostly silted up with a thick layer that hampers smooth water distribution. Other problems included frequent over-irrigation and embankment breaching. On the other hand, tail farms in middle part were poorly or not irrigated, while tail farms in downstream fields were exposed to sand and debris accumulation.

In addition, some 83% of farmers who perceive water supply as extremely uncertain, also experienced irrigation failure on their pre-tilled farms. Of those farmers, about 75% were from the Metateib block where more frequent irrigation failures were observed compared to the Kassala block. All the farmers in the Metateib block perceived that it was less risky to have the first irrigation turn than the second irrigation period. Similar perceptions were shared by 50% of farmers in the Kassala block.

Besides location within the irrigation system, infrastructure (pathways) and water management play an important role in shaping farmers' perception. The poor state of infrastructure and suboptimal O&M of the irrigation network greatly aggravate the risk of insufficient water reaching a farmer's field. Incomplete or lacking canals and control structures combined with poor maintenance limit the capacity of farmers to manage water shortages better and control excess flows. The poor status of irrigation infrastructures aggravates the real and perceived risks and alter decisions about adaptation measures Niles MT and ND Mueller, 2016).

Water users' association. The WUA's main concerns were related to untimely arrival of floods. Late floods may lead to a delayed cropping calendar and attacks by pests and disease, which negatively impact crop yield. This can also give rise to violation of water distribution rules between first and second irrigation fields as well as violation of water rights between upstream and downstream WUAs that are served by the same secondary canal. Violation of water distribution rules between first and second irrigation fields was most prevalent in the downstream Metateib block. Early floods may disrupt ongoing maintenance activities, and lead to water losses on unplanned fields due to infrastructure that is not yet ready to receive water.

Poor water distribution occurs as a result of violation of water-sharing rules between WUAs, and poor control and distribution structures. Without timely and proper maintenance, the irrigation water fails to reach some of the tail fields, due to poor canal transport capacity and distribution structures.

System managers. High floods were the prime concern of water managers, since these events may lead to breaching and severe damage to the main intakes and primary canals. Breaching of river banks and irrigation canals occurs due to weak embankments and poorly operated gates, which allow uncontrolled high floods to enter irrigation system.

Some poor farmers have set up their villages at the tail end of irrigation fields. The breaching may also occur intentionally by these illegal settlers, as they want to avoid irrigation water flooding their homes.

In addition, breaching may occur due to high sediment loads and poor maintenance. But sometimes it is done intentionally by water managers to release water pressure and save the canal system downstream. Similarly, field embankments remain vulnerable to breaching during high flood events, when too much uncontrolled water enters the fields. Because of the lack of proper drainage canals all excess and uncontrolled flow accumulates at the tail of the main canal of the Kassala block, sometimes resulting in consecutive breaching as well as creating waterlogging problems. Similarly, poor maintenance can lead to breaching in the Metateib block.

2.3.5 Consequences

An uncertain water supply influences irrigation system performance and agricultural productivity. Two of the performance indicators used by the GAS managers are the ratio of actual over planned irrigated area and crop yield variability.

Area planned and irrigated. Irrigation performance varies depending on the flood size and location between and within irrigation blocks. The irrigation performance of the downstream block (Metateib) was found to be less adequate compared to the upstream block (Kassala). For Kassala, the ratio of actual over planned area ranged from 68 to 87% during low and high floods, respectively. In contrast, at Metateib, it was 27 and 50% during low and high floods, respectively. The difference could be attributed to location-specific advantages in the system and political bias towards the Kassala irrigation block where more frequent rehabilitation and upgrading works were conducted.

Crop yield variability. Table 2-3 shows the high level of inter-annual variability in sorghum production at the scheme level. Similarly, huge differences were found among individual farmers' fields (yields of 0.2-3.4 Ton/ha). This could be substantiated by the high coefficient of variation (CV), estimated from data of 2010–2013, with values of about 80%.

Table 2-3: Sorghum production in GAS. Data source: GAS records

Year	2004	2005	2006	2007	2008	2009	2010	2011	2012	2013	2014
Crop production (1000 Ton)	19.6	37.3	58.5	42.1	48.6	85.1	43.9	27.5	47.3	40.7	40.6

Moreover, yields were markedly different between the upstream and downstream blocks. The average yield at Kassala (1.4 Ton/ha) was double that of Metateib (0.7 Ton/ha). More than 70% of the farmers at Kassala obtained a good crop yield (>1.5 Ton/ha) during an average flood season, compared to less than 25% at Metateib (Figure 2.6). Nevertheless, the impact of a low flood season on crop yield is similar in both blocks, though only 12% of farmers reported a good harvest. Unfortunately, neither average nor high flood events could promise a better crop yield at Metateib, where more than 70% of farmers harvested less than 1.5 Ton/ha. These differences could be attributed to the downstream location of Metateib, poor infrastructure and system maintenance. These factors also discouraged farmers from adequately investing in crop inputs.

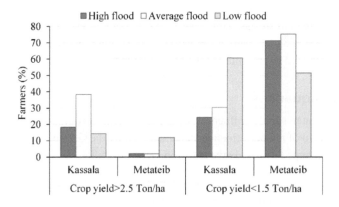

Figure 2.6: Sorghum yield variation at the Kassala and Metateib blocks.

2.3.6 Applying the SPRC model in the spate irrigation setting

In additional to the description given in previous sections, Figure 2.7 and Figure 2.8 provide a summary of the SPRC application in the GAS. The arrows in Figure 7 clearly illustrate the pathways of risk propagation from the river through the primary canals up to the farmers' fields. Additionally, the arrows show the propagation pathway from head, middle to tail farmers, as well as from one field to another adjacent one.

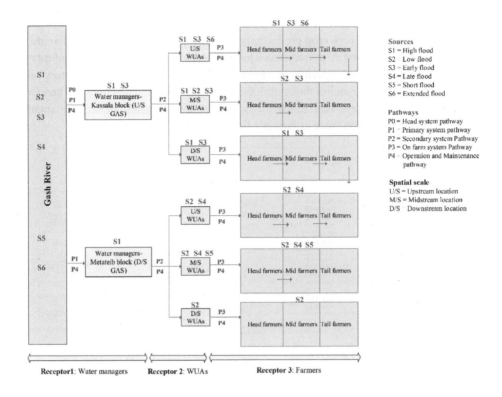

Figure 2.7: Source-Pathway-Receptor model representation in the GAS.

In general, the SPRC model appeared to be a very useful framework for analyzing sources of risks in a spate irrigation system, which originated from the hydrological system, propagated through infrastructure, received by and with consequences for different stakeholders (farmers, WUA and managers). For example, the model was helpful in describing the pathways through which sources of risk were transferred to different parts of the irrigation system, i.e. the Gash River intakes, irrigation structures and canals. Furthermore, the state of O&M was also an important component that could be included in the pathway description. Another noteworthy aspect was its ability to clearly demarcate the actual sources of risk at a detailed spatial level from upstream to downstream locations along the irrigation system, and from head to tail farms along the irrigation fields. On the whole, this SPRC application generated useful insights about system understanding, which could contribute to the formulation of spatially suited plans for interventions.

Apart from these advantages, application of the SPRC framework in the GAS encountered some challenges as well, which could be mainly attributed to cause–effect interchangeability from the point of view of different stakeholders and physical characteristics of the system. For example, the pathways, which are the physical structures, could also be interpreted as physical receptors. From a farmer's perspective, factors such as poor irrigation, mesquite invasion and bank breaching might be considered the sources of risk leading to the poor crop productivity. However, from a system perspective these factors were considered as consequences for the receptors at farm level. This indicates a certain degree of subjectivity in delineating sources, pathways, receptors and consequences, which requires thoughtfulness in interpretation and description. Therefore, the SPRC model seems best suited for larger irrigation systems, rather than smaller components in isolation (such as farmer's fields, or individual blocks).

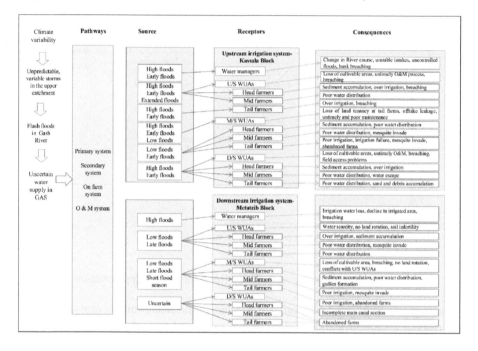

Figure 2.8: The Source Pathway Receptor Consequence model applied in GAS system.

2.4 CONCLUSIONS

This study analyzed irrigation risks and perceptions arising from the uncertain water supply in the GAS—the largest flood-based irrigated area in east Sudan. Using the SPRC framework, this research identified different types of risk elements across this spate irrigation system, their propagation pathways and spatial links, and ultimately consequences for different stakeholder groups (farmers, WUAs and system managers). In general, the SPRC framework proved very useful in analyzing water supply risks, and outlining the different risk perceptions among stakeholders at various locations within the whole system (the GAS). Therefore, its application could be recommended in risk assessment and management studies for flood-based irrigation systems.

The highly variable and unpredictable nature of rainfall in the upper catchment was found to be the primary source of uncertainty. The erratic rainfall in the upper catchments generates unpredictable floods, which, however, are the main source of water for agriculture in the GAS. The whole scheme (e.g. stakeholders and physical infrastructure) faces various consequences from all categories of flood events, 'sources of risks' (late or early, high or low, long or short floods). Moreover, the risks posed by these categories of floods were perceived differently by the stakeholders, and there were varying kinds of consequences such as poor water distribution, poor irrigation, irrigation failure, sedimentation, over-irrigation and abandoned farms. For instance, low floods were the major concern for most farmers, as these events resulted in poor water distribution, and consequently low yields, with most severe impacts on downstream farmers. Untimely floods were the biggest challenge for the WUAs, as different institutional arrangements were needed to cope with them. System managers are greatly concerned by high floods, which can destroy infrastructure, with most consequences in the upstream/head locations. In general, a large variability was evident in crop productivity and overall system performance, with most poor results in the downstream areas.

Moreover, the stakeholders' perception of water supply risks are strongly influenced by their past experiences, which are very much related to the status of the physical infrastructure and O&M strategies. In addition, location factors such as the proximity of irrigation blocks along the river play an important role in how farmers perceive the variability and uncertainty of water supply.

There is a need to acknowledge and strengthen the inherent capabilities of farmers and relevant institutions to identify, assess and cope with the water supply risks arising from hydro-climatic variability. This can be done through capacity-building-oriented programs to increase awareness of risks associated with an uncertain water supply. This study also acknowledged the existence of other irrigation risks related to water governance. Therefore, besides upgrading infrastructure (which is the main focus of existing programs), policy and institutional support could be one of the pillars to increase the capacity to manage risks due to uncertain water supplies. The findings of this research could help formulate mitigation strategies to address the risks faced at different levels of the GAS spate irrigation system in Sudan, but also for similar spate irrigation schemes in arid regions.

3

ADAPTATION STRATEGIES TO COPE WITH LOW, HIGH AND UNTIMELY FLOODS: LESSONS FROM THE GASH SPATE IRRIGATION SYSTEM, SUDAN[2]

In arid areas, water diverted from highly uncertain flash floods is often the only source of water for crop production. Stakeholders in spate irrigation systems have developed numerous measures to cope with uncertain water supply related to low, high and untimely floods. This research evaluates the effectiveness of these measures using the MULINO Decision Support System (mDSS4) tool which is based on the Driving force-Pressure-State-Impact-Response (DPSIR) framework. Using data from interviews with 101 randomly selected farmers, 17 water user associations (WUAs), and 7 system water managers in the Gash spate irrigation system in Sudan, we compare the effectiveness with the rate of adoption. The results reveal the most effective measures are 1) pre-flood preparedness, 2) risk sharing measures through water and land management during and after flood by WUAs, 3) crop management by farmers; and 4) flexibility in operation by water managers. Unfortunately, the most effective measures are not the most adopted ones. The level of adoption is primarily related to the capacity of the farmers, WUAs and water managers to implement the measures without outside support. Generally, measures taken by downstream farmers are less effective than those adopted by upstream farmers due to weak institutional arrangements and lack of adequate resources. Supporting farmers, WUAs and water managers for a wider adoption of the existing effective measures will greatly improve irrigation performance and hence food security in the study area.

---

[2] This chapter has been published in: FADUL, E., MASIH, I. & DE FRAITURE, C. 2019. Adaptation strategies to cope with low, high and untimely floods: Lessons from the Gash spate irrigation system, Sudan. *Agricultural Water Management*, 217, 212-225.

3.1 INTRODUCTION

The adverse impact of climate variability on agricultural production in Sub-Saharan Africa poses a substantial threat to food and water security (Funk *et al.*, 2008). In arid regions of North Eastern Africa (Sudan, Ethiopia, Eritrea), farmers use seasonal flash floods to sustain agricultural production. Flash floods, also called spate, are usually one of the few water sources available in arid areas. Spate irrigation is the practice of diverting flood water from ephemeral rivers to adjacent terraced fields for direct application or into sub-surface storage as soil moisture using simple earthen canals and distribution systems (Steenbergen and Haile, 2010).

Water supply in spate irrigation is highly variable, ranging from large destructive floods to insufficient supplies during the drought years. There is considerable uncertainty in the timing and the volume of floods (Van Steenbergen, 1997). The area that can be irrigated varies each season due to reasons such as the flood volume, the effectiveness of the system to divert floods from highly sediment-laden, unstable rivers, and changes in command area levels (Steenbergen *et al.*, 2010). Other non-climatic factors that influence the spate-irrigated area are the human-environment linkages (Haile *et al.*, 2007) and more broadly the social-ecological interactions (Zimmerer, 2011), which amplify the complexity of the biophysical system.

In general, there is a lack of studies on spate irrigation. A few past studies regarding uncertain water supply in spate irrigation focused on the biophysical and infrastructure-related issues. For example, Van Steenbergen (1997) described how failure of spate irrigation projects in Baluchistan resulted from inappropriate engineering measures such as weak structures, unsuitable design, and wrongly situated structures due to river meandering. The study recommended the adoption of a flexible management approach as well as flexible rules and engineering designs to accommodate variable flows, and an appropriate organisational framework and institutions. Similarly, Khan *et al.* (2014) recommended flexibility of water distribution rules to account for medium and long-term changes in the flood systems of spate irrigation in the Indus River, Pakistan. Haile *et al.* (2008) emphasized the importance of flexibility, not only in view of variations in flow, but also to ensure equitable access and fairness of water sharing.

Past research into spate irrigation studied local experiences with uncertain water supplies primarily in terms of infrastructure, flexibility in engineering design, and water-sharing rules (Saher *et al.*, 2014, Komakech *et al.*, 2011, Ngirazie *et al.*, 2015, Abdelgalil and Bushara, 2018, Haile *et al.*, 2011, Van Steenbergen, 1997, Khan *et al.*, 2014), ignoring farmers' and the water user association's (WUA's) own coping strategies. Effective strategies can greatly reduce the impact of climate variability and enhance local capacity to adjust and cope with the negative consequences (Cooper *et al.*, 2008). Martínez-Alvarez *et al.* (2014) compared farmers' adaptation strategies in normal and dry years at water scarce region in an irrigation district in Spain. The comparison was only limited to strategies dealing with the risks of using of brackish water with high salinity. Ortega-Reig *et al.* (2014) has shown how farmer-managed irrigation system of Valencia in Spain, successfully deal with the river flow fluctuations through a well-established water sharing rules. The study investigated farmers' perceptions on the level of transparency and equity in water sharing, without considering other adaptive strategies for land, soil and crop management, or their effectiveness in reducing the impacts of river flow fluctuations. Similarly, by focusing on irrigation intensity and crop choice strategies, Gaydon *et al.* (2012) found different strategies used during high and low water availability in a farm in Australia's River in a region. Many researchers employed a direct simple approach for evaluating the effectiveness of adaptation strategies using stakeholder's perception (Azumah *et al.*, 2018, Shivakoti and Thapa, 2005, Mcharo, 2013, Pradhan *et al.*, 2017).

From the literature review, we recognized that past studies do not differentiate between the different strategies used for different types of river flow fluctuations, in particular, flood risks such as low, high, and untimely floods. It is also unclear to which extent, and why or why not, these strategies are adopted by the different stakeholders. In addition to closing these gaps, this research added the novel approach of evaluating the strength and weakness of the adaptation strategies to combat risks of uncertain water supply resulting from flash flood diversion. The objectives of this research are: to identify driving force, pressure, state, impact, and response elements of water supply risks; to evaluate the effectiveness of the adaptation strategies; to explore the reasons behind the differences in adoption level by upstream and downstream users; and to evaluate the adoption rate of effective strategies.

To address these objectives, the MULINO Decision Support System (mDSS4) tool based on the Driving force-Pressure-State-Impact-Response (DPSIR) framework was applied to a case study from a spate irrigation system in Sudan, GAS.

3.2 DATA AND METHODS

A mixed-method approach was used for data gathering including: a literature review, field survey, consultations, workshops, and key informant interviews. We applied *the mDSS4 tool* based on the *DPSIR* framework to evaluate the current strategies presenting a logical sequence from: 1) identification of the main elements in the DPSIR framework; 2) identification of the existing adaptation strategies; 3) selection of the evaluation criteria and indicators; and 4) choosing the most effective measures in each strategy that perform better with respect to the selected criteria. The methodology used present a novel approach for evaluation of adaptation measures in a spate irrigation context due to environmental risks and can generally be applied to other irrigation systems with socio-environmental problems characterized by data scarcity.

3.2.1 The mDSS4 tool

The mDSS4 tool, based on the DPSIR conceptual framework, was originally developed in the MULINO project (Multi-sectoral, Integrated and Operational Decision Support System for Sustainable Use of Water resources at the catchment scale). The approach is aimed at facilitating the involvement of stakeholders in the process of integrated water resources management and natural resources management (Giupponi et al., 2004). The mDSS4 tool uses three phases: the conceptual phase, the design phase and the choice phase (Figure 3.1). The conceptual phase involves problem structuring and identification of the study area. The environmental and socio-economic features are linked through cause effect relationships using the DPSIR framework (EEA, 1999). In this context, the Driving forces, represented by natural and social processes, are the underlying causes and origins of pressures on the spate irrigation system. The Pressures are outcomes of the driving forces, which influence the current state of the irrigation scheme. The State reflects the condition/change of the irrigation scheme natural resource, while the Impacts describe the ultimate effects of changes of state. The Responses demonstrate the measures to solve the problems.

The design phase involves: (a) identification of responses (or measures) in terms of the DPSIR framework; (b) selection of a comprehensive set of decision criteria and indicators (Giupponi, 2007); (c) organizing the analysis matrix (AM) which contains the indicator values of the measures for each decision criteria; and (d) building the evaluation matrix (EM) which involves normalization and weighting of the indicator values.

To evaluate the strategies each measure is scored and ranked according to its relevance under each scenario (Sabbaghian et al., 2016) using Multi-criteria decision-making (MCDM), which is widely adopted in irrigation planning (Zarghami, 2011). MCDM shows the impact of different irrigation management decisions (Billib et al., 2009), and evaluates best-management practices for agriculture (Sabbaghian et al., 2016).

The choice phase applies the MCDM to identify the best measures, which perform better on the selected criteria by using decision rule(s) provided in MCDM. Decision rules aggregate partial preferences describing individual criteria into global preferences and rank the measures.

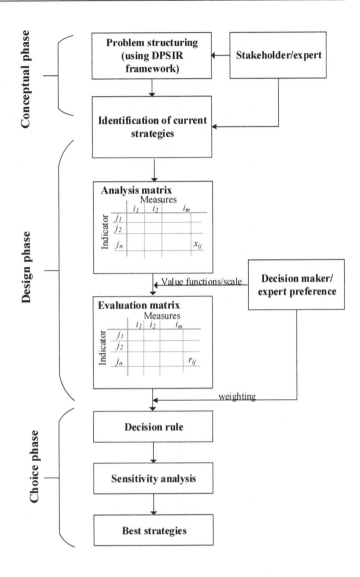

Figure 3.1: Flowchart for the assessment of current strategies using the mDSS4 tool. Source: adapted from the Mullino approach (Giupponi *et al.*, 2004).

3.2.2 Indicator selection and scoring

This study uses qualitative indicators which help to overcome the large uncertainties in the measurement and estimates of variable water supply, and the limited monitoring systems available in the study area. In addition, environmental issues often involve multiple dimension of analysis, uncertainty, and difficulty in obtaining a single measurement scale from different actors (Corral-Quintana *et al.*, 2016). Under such circumstances, the use of qualitative assessment along with the most commonly used indicators in the literature, stakeholder analysis and expert judgment for ranking and prioritization were recommended (de Bruin et al., 2009, Iglesias and Garrote, 2015, Corral-Quintana et al., 2016). The role of farmers and policy makers in interpreting adaptation strategies assist in risk management and decision making (Quiroga and Iglesias, 2009). The suitable criteria and indicators (and associated weights) were defined based on the information gathered during the field visits underpinned by the views of the representative set of stakeholders and experts including researchers, academics, members from related institutes and knowledgeable members of WUAs. The values of the selected indicators ranged from +5 and -5 to reflect the effect of the measure on the elements of the DPSI[3] chain, with 5=very high, 4= high, 3= medium, 2= low, 1= very low, and 0= neutral. The positive and negative signs describe the degree to which the measure positively or negatively affect the criteria. The Simple Average Weighting rule (SAW) was selected to rank the measures (Eq. (1)) based on their performance to meet the indicators.

$$\Phi(a_i) = \sum_{j=1}^{n} w_j \times r_{ij}, \quad \text{for } i = 1, 2, \ldots, m \tag{1}$$

Where: $\Phi(a_i)$ is the overall performance (score) of the ith measure, m is the number of measures, n is the total number of indicators, r_{ij} is the normalized rating of the ith measure with respect to j^{th} indicator and represents an element in the normalized matrix, illustrated by:

$$r_{ij} = \frac{x_{ij}}{max_i x_{ij}}, \quad \text{for the benefit indicator} \tag{2}$$

$$r_{ij} = \frac{1/x_{ij}}{max_i 1/x_{ij}}, \quad \text{for the cost indicator} \tag{3}$$

x_{ij} is an element of the decision matrix which represents the original value of the jth indicator of the ith measure, w_j is the weight of the jth indicator and calculated using the ranking method for assigning weights in order of importance as:

$$w_j = \frac{(n - r_j + 1)^p}{\sum_{k=1}^{n} (n - r_k + 1)^p} \tag{4}$$

Where: p is a parameter for weight distribution; r_j is the rank number of the jth indicator and n is the total number of indicators. The score represents a simple sum of the indicator values of every measure weighted by the vector of weights with $p = 0$. The highest score values are assigned for the best performing measures. Normalization of units was not needed in our case since we are using a similar scale for all the measures. The analysis was implemented for measures to cope with high, low and untimely flood seasons.

---

[3] Driving force-Pressure-State-Impact (DPSI)

A stratified sampling technique was used to collect information from 101 farmers, 17 WUAs, and 7 water managers. The sample was randomly selected from two irrigation blocks located at upstream and downstream along the GAS irrigation project (the Kassala and Metateib blocks). Sample locations was shown in Figure 3.2.

3.3 RESULTS & DISCUSSION

3.1.1 Elements of DPSIR

Following the DPSIR framework, the cause-effect links from human environment problems in the study area (Figure 3.3), were briefly conceptualized as follows: Driving forces consist of climate change and variability, floods & droughts, population growth, poverty, changing laws, policies and regulations that support spate irrigation institutions. Pressures include increased diversion of highly sediment-laden floods, introduction of mesquite to tackle droughts, lack of clear decision making in the system operation, ambiguous institutional arrangements and enforcement. State is represented by uncertain and inequitable water supply, sedimentation of irrigation infrastructures and fields, increased rate of mesquite spread in fields, river banks & irrigation canals, poorly maintained and damaged infrastructure. Impacts result in abandoned fields as a result of irrigation failure, reduction in the efficiency of irrigation supply and frequent damage to infrastructure, decrease in crop productivity, lack of organized marketing. Response describes the measures and strategies reported by the farmers, WUAs and water managers to influence, reduce or mitigate any element of the Driving force-Pressure-State-Impact (DPSI) chain.

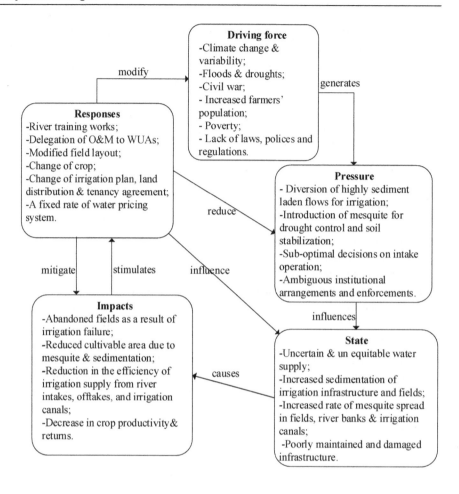

Figure 3.2: Driving force Pressure State Impact Response framework applied to the Gash Spate Irrigation Scheme Source: adapted from the Mullino approach (Giupponi et al., 2004).

3.1.2 Identification of measures against uncertain water supply

The unpredictability of floods leads to uncertain and unequitable irrigation water supply due to low, high, and untimely flows into the irrigation system. Figure 3.4 illustrates the daily variability of the Gash River flows measured at the Kassala Bridge and at the Salam-Alikum irrigation intake. We identified, a series of measures performed before, during and after the flood season that can accommodate different uncertain floods (Table 3-1). Additionally, we coded and grouped different measures, adapted by different stakeholders, based on their relevance to the emerging flood risk e.g. low, high and untimely floods (Table 3-2, Table 3-3, Table 3-4).

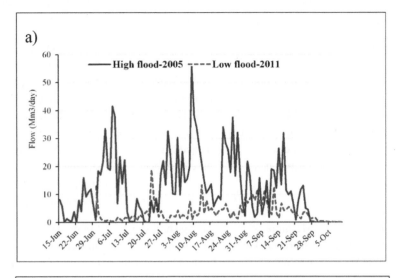

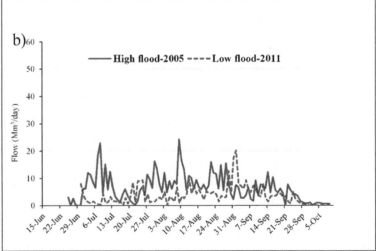

Figure 3.3: Daily hydrographs at: a) Gash River, b) Salam-Alikum intake. Source: Gash River Training Unit.

Table 3-1: Measures (responses) to cope with uncertain water supply

Stakeholder	Before flood measures	During flood measures	After flood measures
Farmers	Land preparation before flood, Use of shrubs and weeds, Pre-tillage practice, Make small earth bunds	Use of shrubs and weeds, Digging small ditches to distribute water flow, Use lebsha to reduce velocity, use of sand bags to close breaches	Sharecropping, Cultivate vegetables, Increase seeding rate for fodder production, Double tillage, Wetting seeds, Reduction of cultivated area, Change crop variety, Social system of sharing benefits, Delay cropping date, Change crop, Do not cultivate.
Water User Associations (WUAs)	Mesquite clearance, Land leasing, Pumping Groundwater, Fixed land system, Re-alignment of field canal	Laying shrubs downstream field inlets, Flood water spreading at fields, Monitoring breaching events, Embankment heightening, Manage irrigation period between WUAs	Lottery system for field allocation, Change field-spur location if needed
Water managers	Embankments heightening, Share some of maintenance activities with WUAs, Flexible irrigation plan	Flexibility in water allocation period, Close monitoring of flooded areas, Manual control of intake diversion, Delay of maintenance of inaccessible areas	Water pricing based on actual irrigated area at affixed rate per irrigation unit (Feddan)[4]

---

[4] Feddan =4200 square-meters

Table 3-2: Low flood strategy for different stakeholders

Farmers' measures	Code
Reduction of cultivated area	SF-Low1
Pre-tillage before flood season	SF-Low2
Land preparation before flood	SF-Low3
Double tillage	SF-Low4
Increase seeding rate for fodder production	SF-Low5
Wetting seeds to reduce the 1st growth developing stage	SF-Low6
Make small earth bunds	SF-Low7
Sharecropping	SF-Low8
Change sorghum variety	SF-Low9
Cultivate vegetables	SF-Low10
Summer tillage	SF-Low11
Exit cropping season (Do not cultivate)	SF-Low12
Digging small ditches to distribute water flow	SF-Low13
Use of shrubs and weeds	SF-Low14
Cultivate only on part of the field	SF-Low15
Social system of sharing benefits	SF-Low16
WUAs' measures	**Code**
Lottery system for field allocation to farmers	SWUA-Low1
Fixed system for field allocation to farmers	SWUA-Low2
Flexible infield spurs for field water distribution	SWUA-Low3
Manage irrigation period between adjacent WUAs	SWUA-Low4
Re-alignment of field canal	SWUA-Low5
Allocation of one farm per farmer in every flooding season	SWUA-Low6
Mesquite clearance	SWUA-Low7
Participate in flood water spreading	SWUA-Low8
Sharing field canal between adjacent fields on different irrigation time	SWUA-Low9
Laying shrubs and weeds at field head to dissipate flow energy	SWUA-Low10
Change of water source to groundwater pumping at head fields	SWUA-Low11
Longitudinal field division of irrigation fields	SWUA-Low12
Temporarily land leasing to private sector	SWUA-Low13
WUAs split in subgroups of farmers	SWUA-Low14
Water managers' measures	**Code**
Division of flood period in two irrigation schedules	SM-Low1
Allocation of fields with high and low chances of good irrigation for each WUA	SM-Low2
Mapping of flooded areas every 10 days	SM-Low3
Water allocation period with flexibility	SM-Low4
Share maintenance burden with WUAs	SM-Low5
Diversion of first floods to groundwater recharge and drinking basins	SM-Low6

Table 3-3: High flood strategy for different stakeholders

Farmers' measures	Code
Use of sand bags for small breaches	SF-High1
Use of lebsha to dissipate flow energy D/S field intakes	SF-High2
Fill the breach with shrubs and weeds(Lebsha)	SF-High3
Delaying the start time of cropping activities	SF-High4
Change crop variety	SF-High5
Close of water paths and gullies	SF-High6
Breaching banks of nearby fields	SF-High7
Cultivate water melon in winter	SF-High8
Double tillage to reduce weeding	SF-High9
Cultivation of a second crop after harvest	SF-High10
WUAs' measures	Code
Field preparation (field canal desilting, heightening embankments)	SWUA-High1
Laying shrubs and weeds at field head to dissipate flow energy	SWUA-High2
Breaching embankments of adjacent fields	SWUA-High3
Report major breaching	SWUA-High4
Water manager' measures	Code
River training works and strengthening of River embankments	SM-High1
Routine maintenance before flood season	SM-High2
Maintaining critical sections before flood	SM-High3
Maintaining a reasonable distance between field offtakes	SM-High4
Cooperation with River monitoring units for early warning	SM-High5
Established water level gauges at intakes for effective operation	SM-High6
Start irrigation with first upstream and last downstream fields	SM-High7
Flow releases to planned fields if canal stability is not threatened	SM-High8
Diversion of water into unplanned fields to release flow energy	SM-High9
Delay of maintenance work to the end of season	SM-High10
Raising offtakes of main canal and secondary canal	SM-High11
Use of labour to prevent accumulation of debris U/S offtakes	SM-High12
Mobilizing financial resources and incentive system	SM-High13

Table 3-4: Untimely flood strategy for different stakeholders

Farmers' measures	*Code*
Use of sand bags for small breaches & seek assistance for major breaching	SF-untimely1
Fill the breach with shrubs and weeds (Lebsha)	SF-untimely2
Change crop	SF-untimely3
Exit cropping season (Do not cultivate)	SF-untimely4
Social system of sharing benefits by sharing irrigated fields or harvest	SF-untimely5
Cultivate in winter	SF-untimely6
WUAs' measures	*Code*
Manage irrigation period between adjacent WUAs	SWUA-untimely1
Lottery system for field allocation to farmers	SWUA-untimely2
Use sand bags and seek assistance	SWUA-untimely3
Water managers' measures	*Code*
Maintaining critical sections before flood,	SM-untimely1
Emergency action on silt removal	SM-untimely2
Borrow maintenance equipment where possible	SM-untimely3
Use of timber stop logs to control water level	SM-untimely4
Use of experienced gate operator to adjust openings	SM-untimely5
Priority of maintenance to WUAs who paid water fees	SM-untimely6
Borrow from state government to deal with delay of budgetary flow	SM-untimely7
Involve private sector for maintenance activities	SM-untimely8
Allowing flexible starting and end dates of irrigation	SM-untimely9

3.1.3 Evaluation of measures for the adopted strategies

Using the mDSS4 tool, the reported measures used for coping with low, high and untimely floods were assessed on identified environmental, management, social and economic criteria (Table 3-5). Results of the top-ranking measures and scores are presented in Table 3-6. A complete list of the measures and scores is provided in Appendix 3.I.

To measure the effectiveness, we used performance scores; ≥ 0.7 signifies a high effectiveness, scores between 0.5 and 0.7 indicate medium effectiveness, and scores<0.5 show least or no effectiveness. Table 3-7 lists the number of measures developed with their effectiveness in reducing or controlling the elements of the DPSI.

The analysis reveals striking differences in effectiveness scores according to the location and type of stakeholder, as discussed below.

Table 3-5: Indicators selected for the analysis

Criteria	Indicators	Position in DPSIR	Decision maker	Definition
Management	Flexibility	Driving	All	Measures that can be delayed, abandoned, expanded or contracted
	Robustness	Driving	All	Measures that can operate under different conditions
	Technical-difficulty	Pressure	All	Measures that can be implemented with the available technical, managerial and financial capacity
	Effective & efficient operation of irrigation systems	Pressure	Water manger, WUAs	Measures that enhance operation of offtakes, water distribution structures and intakes
	Effective & efficient soil water management	State	Farmer	Measures that improve soil water management and soil infiltration
	Effective & efficient diversion systems	State	Water manager	Measures that improve the diversion structures or irrigation water from river intakes
	Effective & efficient maintenance of irrigation systems	Impact	Water manger, WUAs	Measures that enhance maintenance of irrigation system such as canals, structures, fields, desilting
	Effective & efficient field water management	Impact	WUAs, farmers	Measures enhance performance of field infrastructures such as field canals, offtakes, spurs, field embankments, field slope, etc.
Social	Equity	State	All	Measures that enhance equitable access to land and water and other resources
	Social acceptance	Driving	All	Measures that are acceptable by all stakeholders e.g., farmer, WUAs, water managers

Environment	Invasive weed control & management	State	All	Measures contributes to control or manage mesquites in irrigation system
	Sediment reduction & management	State	All	Measures that minimize or manage sediment in the irrigation system
Economic	Mobilization of financial resources	State	Water manager	Effective and efficient service pricing and incentive system
	Physical productivity	Impact	Farmer, WUAs	Measures that enhance crop yield productivity
	Economic productivity	Impact	Farmer	Measures that enhance crop profitability

Note: WUAs denotes water user associations

Table 3-6: The adopted measures and their scores

Stakeholder	Low flood strategy	Score	High flood strategy	Score	Untimely flood strategy	Score
Farmers	Land preparation before flood	0.81	Use of lebsha to dissipate flow energy downstream field intakes	0.84	Use of sand bags for small breaches & seek assistance for major breaching	0.91
	Sharecropping	0.74	Use of sand bags for small breaches	0.72	Cultivate in winter	0.72
	Use of shrubs and weeds	0.70	Cultivate water melon in winter	0.71	Fill the breach with shrubs and weeds	0.59
	Summer tillage	0.65	Delaying the start time of cropping activities	0.64	Change crop	0.54
WUAs	Mesquite clearance	0.72	Field preparation (field canal desilting, heightening embankments, etc.)	0.91	Lottery system for field allocation to farmers	0.80
	Temporarily land leasing to private sector	0.72	Report major breaching	0.80	Manage irrigation period between adjacent WUAs	0.45
	Laying shrubs and weeds at field head to	0.71	Laying shrubs and weeds at field head to	0.76	Use sand bags and seek assistance	0.05

	dissipate flow energy		dissipate flow energy					
	Change of water source to groundwater pumping at head fields	0.71	Breaching embankments of adjacent fields	0.09			
Water managers	Mapping of flooded areas every 10 days	0.68	Routine maintenance before flood season	0.84	Involve private sector for maintenance activities	0.83
	Diversion of first floods to groundwater recharge and drinking basins	0.59	A fixed rate for water pricing and incentive system	0.82	Allowing flexible starting and end dates of irrigation	0.63
	Allocation of fields with high and low chances of good irrigation for each WUA	0.51	Raising offtakes of main canal and secondary canal	0.75	Use of timber stop logs to control water level	0.58
	Water allocation period with flexibility	0.50	Established water level gauges at intakes for effective operation	0.71	Priority of maintenance to WUAs who paid water fees	0.57

Note: WUAs denotes water user associations

3.3.3.1. Difference between farmers, WUAs and water managers

The farmers and WUAs developed a larger number of medium and highly effective measures for low flood strategy than water managers. Meanwhile, water managers developed more effective measures to cope with high and untimely floods; 89% and 85% of their measures were medium to highly effective, respectively. Further, WUAs were comparatively poor in developing effective measures for high and untimely floods unlike individual farmers at their farms. Yet the majority of farmers' measures were not among the highly effective ones. These results were largely influenced by the level of perception towards flood risks (Fadul *et al.*, 2018, Niles and Mueller, 2016). Performance of the measures used by farmers, WUAs and water managers is shown in Fig. 3.4a-c, respectively.

Table 3-7: Number and effectiveness of the measures used for each flood strategy

Stake holder	Low flood strategy			High flood strategy			Untimely flood strategy		
					Score				
	≥ 0.7	<0.7&$\geq$ 0.5	< 0.5	≥ 0.7	<0.7&$\geq$ 0.5	< 0.5	≥ 0.7	<0.7&$\geq$ 0.5	< 0.5

Farmers	3	7	6	3	6	1	2	3	1
WUAs	5	6	3	3	0	1	1	0	2
Water managers	0	4	2	7	4	2	1	7	1

Note: WUAs denotes water user associations

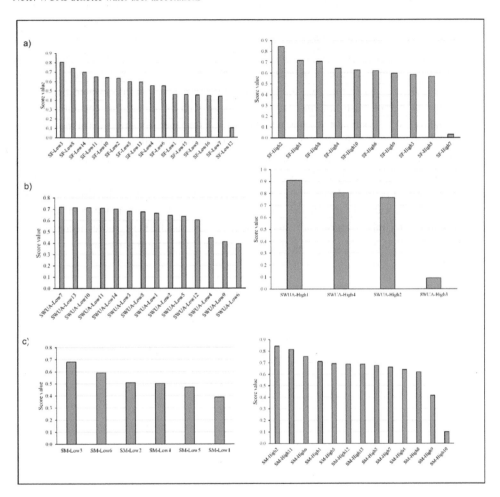

Figure 3.4 a, b, c: Performance of low and high flood strategies, a) at farmer level, b) at WUA level, c) at water manager level.

Note: SF-Low, SF-High denotes farmers 'measures for high and low floods; SWUA-Low and SWUA-High are the water user associations' measures for low and high floods; SM-Low and SM-High are water managers' measures for low and high floods. The codes of individual measures and scores are explained in Appendix 3.1.1-2.

3.3.3.2. Difference between upstream and downstream farmers

There is a clear difference between measures taken by farmers in upstream and downstream blocks. The differences can be attributed to the location and performance of the WUAs and water managers. In the upstream Kassala Block, improved field water distribution systems (such as infield canals and spurs to deflect water) are more prevalent than in the downstream Metateib block (Figure 3.5a-c). Additionally, farmers in the Kassala block are collectively organized in subgroups to operate and maintain the large sized fields, unlike farmers at Metateib block who work individually, hence reduced cultivable area per farmer (0.1-0.2 ha). The collective action in flood water distribution, access to agreed farm holding size (1.25 ha), as well as the conflict resolution by the organized WUAs supported wider adoption of risk management strategies at the upstream block than the downstream block. Being the first irrigation block to divert flood water, the Kassala block could suffer from the impact of potentially destructive high floods but also has the opportunity to divert water during the whole season, unlike the downstream block which receives floods in lesser quantities. The proximity of the Kassala block to the Kassala city has drawn the attention of donors, GAS administration and more qualified technical staff and labour which has resulted in more technical and administrative support to WUAs and hence, better O&M of infrastructure.

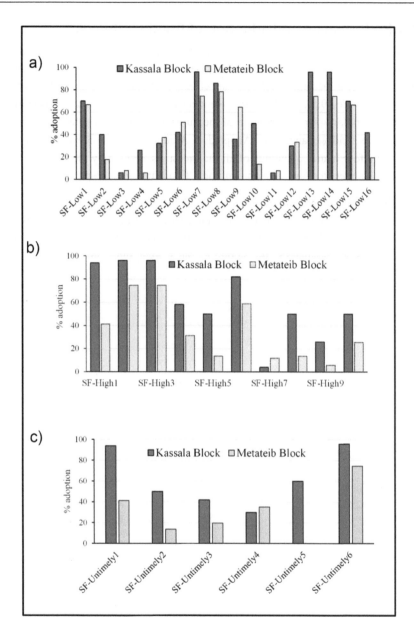

Figure 3.5 a, b, c: Comparison of % adoption of flood strategies used by farmers at upstream and downstream block during: a) low floods, b) high floods, c) untimely floods.

Note: SF-Low, SF-High, SF-Untimely are the farmers' measures taken for low, high and untimely floods.

3.1.4 Performance and adoption of flood strategies

After assessing the effectiveness of coping strategies with different flood types, we compared them with adoption rates as reported in interviews during our study. For simplicity, we grouped high and medium effective measures in Highly Ranked (HR-performance score ≥0.5), and low effective measures in Low Ranked (LR-performance score<0.5). Similarly for the adoption of measures, we selected: Highly Adopted (HA-adoption rate ≥ 50%), and Low Adopted (LA-adoption rates<50%). Figure 3.6a-c illustrate the four matrix quadrants of all the measures adopted by the farmers, WUAs, and water managers for low, high and untimely flood strategies, respectively.

The analysis revealed that all HA measures could be implemented by farmers and WUAs without external support or technical difficulty. However, most HR measures were not well adopted by all the stakeholders: for the HR measures in the case of the farmers, WUAs, and water managers, the adoption rates were only 38%, 11%; and 29%, respectively.

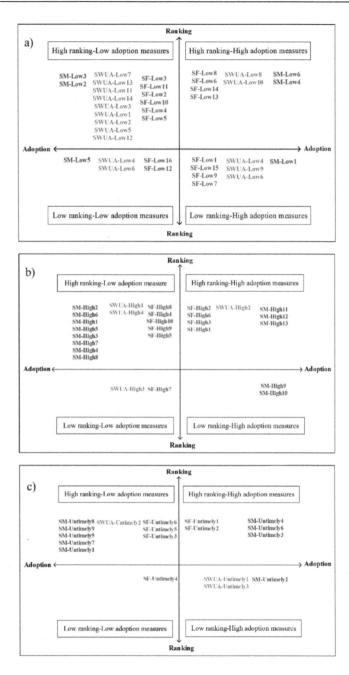

Figure 3.6 a, b, c: Performance and adoption of flood strategies: a) low flood, b) high flood, c) untimely flood.

Note: SF-Low, SF-High , SF-Untimely denote farmers' measures for low, high and untimely floods; SWUA-Low, SWUA-High and SWUA-Untimely denote water user associations' measures for low , high and untimely floods; SM-Low ,SM-High and SM-Untimely denote water managers' measures for low, high and untimely floods. The codes for individual measures and scores are explained in Appendix 3.I.1-3.

3.3.4.1 Low flood strategy (Figure 3.6a)

The most effective measures to cope with low floods (and hence low water availability) are performed before the start of the irrigation season such as land and soil preparations (farmers), and mesquite clearance (WUAs). For water managers the most effective measure is frequent monitoring of the irrigated area, which allows for timely termination of irrigation if fields have been supplied with enough water. The lottery system for field allocation, based on distributing the irrigable area, scored similarly to a fixed system, which gives the farmer the right to access the same part of a particular field all the time. Fixing the location of a tenancy was implemented in 2004 with the introduction of the WUAs' Act to allow farmers to clean and prepare their own part of farms. Sharecropping can adjust to low flood risks such as irrigation failure, or poorly irrigated field. Increasing income by changing the cropping plan from sorghum to fodder/vegetable production is a flexible cropping activity that can at least provide fodder for their livestock.

A large number of HR measures are not adopted by farmers and WUAs. The adoption level depends on the capacity, available resources and support systems. For example, field preparation and mesquite clearance needs financial and institutional support, which are not available to the majority of farmers and WUAs. Therefore, HR measures that could be adopted by the farmers without outside support include sharecropping, wetting seeds, and making in situ flood harvesting. Similarly, WUAs could easily adopt HR measures developed for uniform water distribution e.g. water spreading, use of shrubs & weeds at field head. The results for water managers were influenced by the differences between the upstream and downstream block. In the downstream Metateib block HR measures such as frequent monitoring of the flood area and water allocation rules allowing for fallow fields are not practiced, while LR measures (such as options to minimize the losses of a low flood season with the least possible benefits) show high adoption rates. Finding opportunities with even LR measures is a way to deal with the flood risks when no other options are available.

3.3.4.2 High flood strategy (Figure 3.6b)

Simple traditional methods with less dependency on external support were among the measures most adopted by farmers. The promising measures include delaying the date of cropping to reduce soil moisture and increasing income from cash crop production such as water melons. However, these measures have not been widely adopted due to the lack of investment needed until harvest. The best measures for WUAs were desilting of field canals and secondary system, and raising embankments.

Maintenance activities are poorly adopted by WUAs due to poor access to equipment and resources. Unlike WUAs and farmers, water managers were better equipped for dealing with the high floods. Almost all the measures adopted by water managers were ranked as medium or highly effective in dealing with high floods. Nevertheless, adoption of measures was location-dependent. As there is more concern of high to divert floods, developed and adopted most of the best scored measures compared to the downstream Metateib block. Nevertheless, the best scored measure (routine maintenance of the irrigation system) is less adopted in both blocks due to the limited amount of equipment and human and financial resources.

3.3.4.3 Untimely flood strategy (Figure 3.6c)

There is a relatively low number of measures dealing with untimely floods (early, late, long and short floods). If untimely floods occur during the maintenance and preparation activities, this may lead to breaching of the poorly maintained canals and structures. Farmers' most effective measure is the use of sand bags and locally made lebsha (i.e. vertically laid sticks filled with dry shrubs); this accumulates the sediment from the flow to form a cementing material to reduce flow energy. Figure 3.8 illustrates a lebsha installed at a field entrance. Although changing the crop and delaying cropping were highly scored, farmers were reluctant to change their crop due to their preference for subsistence crop production and the high cost of cash crop production.

The lottery system was the HR measure for the WUAs to distribute the actual irrigated area between farmers. WUAs in Metateib have heavily adopted the lottery system as a result of the high risk of poor irrigation in the downstream block.

Involving private companies in the maintenance work, to prepare irrigation system to accommodate the untimely floods, was the best measure for water managers and, in particular, highly adopted in the Kassala block. Figure 3.9 demonstrates the impact of untimely flood occurred during maintenance activities. Similarly, flexibility in the start and end dates of irrigation were highly adopted in the Kassala block as compared to the Metateib block where conflicts and disagreement between WUAs are more prevalent.

In general, low adoption of HR measures were observed at all levels. While LR measures such as adopting an emergency maintenance plan for the most critical section, infringement of water rights of downstream WUAs and exit cropping were heavily adopted as the minimum action to reduce the losses by water managers, WUAs and farmers, respectively.

Figure 3.7: The use of lebsha at a field entrance.

Figure 3.8: An untimely flood occurred during maintenance activities in season 2014.

3.4 CONCLUSION

The stakeholders at all levels in the Gash agricultural scheme (GAS) in Eastern Sudan had developed numerous measures to cope with uncertain water supply, ranging from too low, too high and untimely floods. This study evaluated the effectiveness of these measures adopted by the farmers, water user associations (WUAs) and system water managers. The research used the Driver-Pressure-State-Impact-Response (DPSIR) concept for problem structuring and the mDSS4 software for the evaluation, based on a set of selected criteria and indicators most relevant to the local context of spate irrigated agriculture.

The evaluation of coping strategies revealed a low correlation between effectiveness and adoption. Most of the highly (effective) ranked measures had a low adoption rate by farmers, WUAs and water managers. Similarly, some of the poorly ranking measures were among the most frequently adopted ones. The most effective measures adopted by the farmers to cope with risks of low floods were characterized by activities performed before flood mostly for field and soil preparations. Generally, frequently adopted measures were characterized by low dependency on external resources. There were limited measures developed by the WUAs to deal with high and untimely floods due to inaccessible resources. Water managers have developed effective strategies for high and untimely floods, however effective strategies are well practiced at the upstream block and very poorly practiced at the downstream block.

The promotion of highly effective but less frequently adopted measures could boost the productivity of the GAS. In this regard, there is a need for institutional arrangements with respect to operation and maintenance, building capacity of the water managers and WUAs to strengthen their roles and capabilities for taking effective actions. Additionally, the adopted measures should be strengthened with better knowledge and research with respect to crop water use, the optimum irrigation schedule, crop yield response to different flood scenarios and management options.

The application of DPSIR framework in the spate irrigation context has been useful in problem identification using a holistic approach. This assists in understanding and focusing on the element that need to be improved. In this study, the pressure element described by the technical difficulty (including technical, financial and management capabilities) indicator was found to be the most sensitive factor that can introduce change towards improved spate water management. This is in line with the findings by Bashier *et al.* (2014)who assessed the performance of WUAs in GAS and concluded that WUAs in GAS were technically and financially poor at managing spate water systems.

While much knowledge still needs to be known about complex spate systems, it is vital for farmers and WUAs to participate in the planning, implementation, and operation to have productive spate systems.

To improve performance of water management in spate irrigation, adoption of effective strategies to cope with risks related to uncertain water supply is crucial. We observe that the most effective measures are least adopted and vice versa. Most of the effective measures are constrained by: 1) lack of institutional, policy and technological support for irrigation, 2) limited access to irrigation water, and 3) limited access to maintenance services.

For *farmers,* land preparation, soil and crop management, and improving equal access to well irrigated fields can be enhanced through revisiting land rights at farm level and provision of credits. These efforts could be useful to reduce and mitigate risks faced by the farmers. For *WUAs,* programs for the eradication of mesquite, involvement of private sector for field levelling and preparation, reducing the irrigation failure of large sized fields by improving field design, revisiting regulation of water rights field offtakes, access to maintenance equipment and provisional of financial resources can increase the irrigation efficiency and equity at WUA field level. For *water managers*, provision of timely service to WUAs to maintain the field and secondary system can result in efficient operation and maintenance. Efficient operation of intakes and distribution systems can be achieved through: enhanced policy and institutional set-up of GAS administration, upgrading technical staff to ensure a continuous operation and maintenance process before, during and after the flood season, and provision of WUAs with access to equipment and capacity building. These efforts could be useful in reducing and mitigating risks faced by the farmers, WUAs and water managers and hence improve productivity and sustainability.

4

IRRIGATION PERFORMANCE UNDER ALTERNATIVE FIELD DESIGNS IN A SPATE IRRIGATION SYSTEM WITH LARGE FIELD DIMENSIONS[5]

The sustainability of spate-irrigated agriculture in a semi-arid climate depends on efficient use of irrigation water. Thus, efficient capture and storage of soil moisture in the field are crucial for sustained productivity. The main objective of this study is to examine the performance of improved field design strategies to manage variable irrigation water supply and application time in the Gash agricultural scheme (GAS) in eastern Sudan where open-end border irrigation is practiced to irrigate large fields with variable sizes that range from 250 to 1,250 ha. Irrigation performance was examined using the WinSRFR model for a large-sized field (8,400 m×500 m), continuously irrigated for 25 days but also under alternative designs and irrigation times. The performance was evaluated using efficiency, adequacy and uniformity criteria. The results demonstrate that the current irrigation practices are quite inefficient but could be substantially improved by adopting alternative design and operational strategies. A vertical division of the field (8,400 m ×250 m) under the average inflow condition could result in a substantial increase in application efficiency (from less than 50% to over 70%), distribution uniformity (from 0.34 to 0.87), and irrigation adequacy (from 0.68 to 1). Additionally, the fields could be irrigated in considerably less time when an alternate irrigation schedule between two equally divided fields is followed, which indicated time savings of 40% under a high inflow rate scenario (occurring during a large flood season), and a 20% reduction in time under an average inflow rate scenario (occurring during a medium flood season).

[5] This chapter is based on a paper submitted to the Journal of Agricultural Water Management in: FADUL, E., MASIH, I., DE FRAITURE, C. & SURYADI, F. X. 2019. Irrigation performance under alternative field designs in a spate irrigation system with large field dimensions.

4.1 INTRODUCTION

Sudan is among the countries most vulnerable to climate change and variability in the world (USAID, 2016). Recently, GIEWS (2018) estimated 6.2 million people are vulnerable to severe localized food insecurity due to conflict and weather shocks. Food insecurity in Sudan is directly linked to climatic and non-climatic factors, including climate change & variability (Osman-Elasha *et al.*, 2006), conflicts & internal displacement of the population (Gundersen, 2016), uncertainty in agricultural production (Muli *et al.*, 2018) and low crop productivity (Siddig and Babiker, 2012). Climate change is also predicted to have the potential to alter natural flow regimes (Palmer *et al.*, 2009) including intermittent rivers, such as the Gash River in eastern Sudan. This river is the main source of water supply for a major spate-irrigated scheme called the Gash Agricultural Scheme (GAS). GAS provides a source of livelihood for a large portion of the population (almost 500,000) in eastern Sudan. Most of the population in the scheme area live in poor communities whose livelihood conditions are threatened by the climate-based farming systems and increased food insecurity.

Spate irrigation is the practice of diverting flash floods from intermittent (ephemeral) river beds through irrigation canals to fields surrounded by earthen bunds (Lawrence and Van Steenbergen, 2005) whereby the large volume of flood water induced by precipitation in the upper catchment is directed to low land and wadi areas (Haile *et al.*, 2006). Unlike conventional irrigation which relies on less variable supply from perennial rivers, spate irrigation relies on highly variable water supply (Van Steenbergen, 1997). In GAS, water supply variability induces uncertainty in farming decisions due to unpredictable floods in terms of timing, volume and frequency (Haile *et al.*, 2011, Fadul *et al.*, 2018). Variability exerts high pressure on the operational decisions at the intakes of the GAS irrigation scheme and on farming decisions. The variability is illustrated by flood size (large, medium or small floods), flood timing (early or late floods), and flood duration (short or extended floods) (Fadul *et al.*, 2018). Farmers have to deal with the large uncertainty and variability through coping strategies characterized by special arrangements for crop, land and water management that have increased their resilience to climate threats (Osman-Elasha *et al.*, 2006, Fadul *et al.*, 2018). For example, there are special rules for water and land rotation, field design, field size, and crop choice among others.

Field designs and field water management have been developed to deal with the uncertainty using large border fields and one-off water application (25-28 days) before a crop is planted. However, as the cropping pattern has changed and the competition for water increased, farmers have started complaining of poor crop yield, poor irrigation adequacy and poor water distribution (Fadul *et al.*, 2018).

Therefore, in view of the changed water demands we evaluated the field water distribution efficiency and adequacy under the prevailing practice of large fields with open-end borders. We also explored experiments with adapted field layouts and water application strategies that farmers can adopt on a small scale to improve water distribution efficiency and irrigation adequacy.

With limited literature on spate irrigation, field water management was described in terms of water rights and rules, field structures, field distribution systems and size of command area (Steenbergen *et al.*, 2010, Komakech *et al.*, 2011, Van Steenbergen, 1997, Haile *et al.*, 2011, Koppen *et al.*, 2007). Previous studies on field design and optimization focused on conventional irrigation systems with a fixed predetermined supply and application time (Bautista *et al.*, 2009b, Zerihun *et al.*, 2005, Salahou *et al.*, 2018, Adamala *et al.*, 2014, Anwar *et al.*, 2016, Bo *et al.*, 2012) while research on field design in spate irrigation characterized by water supply variability is uncommon. Moreover, research on optimizing field dimensions and cut-off times (T_{co}) for a range of unpredictable inflow volumes under the complex conditions encountered in spate irrigation are lacking.

This research is focused on farm improvements that consider field design and field water operation practice to manage water supply variability during large, medium and small floods in GAS. We apply the hydraulic simulation model WinSRFR 4.3.1 to examine the performance of field irrigation using application efficiency (AE), distribution uniformity (DU), and adequacy (AD) criteria. The specific objectives of this research are: (1) to determine the optimum field dimensions and cut-off time/application time (T_{co}) which satisfy the required irrigation depth; (2) to establish different scenarios of T_{co} at different inflow rates (Q) representing high, medium and small flood seasons; and (3) to assess the sensitivity of the selected optimum field dimensions to changes in design parameters.

4.2 MATERIALS AND METHODS

4.2.1 Field water management

The border irrigation method is the most dominant technology for field water management in spate irrigation systems. This method is suitable for large-scale farming characterized by long uniformly graded strips and separated by earth bunds (FAO, 1988). The bunds direct/divert large volumes of water flows towards the downstream end over a relatively flat sloped area in a short time. This irrigation method suits the production of many food crops which is an important element for food security in poor marginal areas that are sustained by spate irrigation in developing countries.

In GAS, water is released from canal offtakes and allowed to pass freely from the upstream head to the downstream open-end border fields with variable sizes (250-1,250 ha), as shown in Table 4-1. Substantial variability in the flow rate entering irrigation fields is common (Figure 4.1).

Currently, GAS operates on a two-yearly rotation system in which the first half of the irrigation fields is irrigated in one season and the second half is left fallow to be irrigated the next season leaving the first half as fallow. The effective irrigation period extends for 60-70 days from July to September, divided into two periods. In the first irrigation period, the water is passed through field offtakes to irrigate 60% of the annual planned area. The second irrigation period covers the remaining 40% of the planned area for irrigation in that season. In the first irrigation period, field offtakes supply irrigation water continuously for a period of 25-28 days. The second irrigation period starts after the first irrigation is completed. Water is applied in a one-off irrigation approach with no further irrigation. Amarnath *et al.* (2018) studied water consumption in GAS using smart ICT to provide weather and water information to smallholders and found that with the farmers' practice of one-off irrigation application, no further irrigation was needed as the soil was well irrigated.

Water movement follows the land slope and/or is directed by a system of field spurs carefully distributed along one or both sides of field the embankments. Three types of field layout systems are distinguished in GAS (Figure 4.2): Type A) fields supported by field canals and deflecting spurs, Type B) fields without field canals and spurs, and Type C) vertically divided fields. Shortcomings of type A are the frequent maintenance requirements of field spurs such as rebuilding or relocating. Type B is exposed to gully formation and require frequent land levelling using machinery which are often inaccessible (Figure 4.3). The longitudinally divided field, Type C, is rarely found and based on field trials without scientific justification or tests.

Table 4-1: Irrigation fields in GAS

Field area (ha)		250-500	>500-750	>750-1,000	>1,000	Number of fields
Irrigation block	Kassala	22	52	22	4	27
	Mekali	70	30	0	0	48
	Degain	88	11	0	1	35
	Tendeli	31	42	27	0	38
	Metateib	50	41	9	0	36
	Hadaliya	47	53	0	0	28
% of total fields		53	36	10	1	

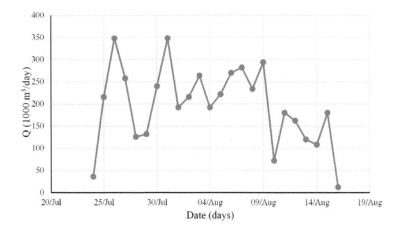

Figure 4.1: Flow hydrograph of an irrigation field in Kassala irrigation block during the 2015 flood season. Source (HRC, 2016).

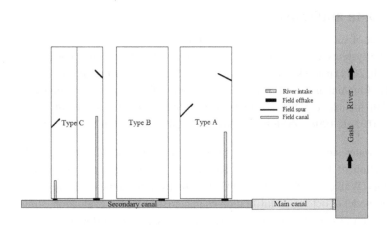

Figure 4.2: Layout of irrigation fields in GAS.

Figure 4.3: Field spur used for redirecting water.

4.2.2 Data and tools

A number of surface irrigation simulation models have been developed to assess border irrigation systems and minimize the complexity of design computations (Adamala *et al.*, 2014, Ali, 2011) using kinematic waves, hydrodynamics, zero inertia and volume balance models (Ebrahimian and Liaghat, 2011). Some of the simulation models for border irrigation include BRDRFLW (Strelkoff, 1985), BORDER (Strelkoff, 1990), SRFR (Strelkoff *et al.*, 1998), ZIMOD (Abbasi *et al.*, 2003), and SIDES (Adamala *et al.*, 2014).

Surface water simulation software, such as SIRMOD, SISCO and WinSFR, perform optimization of field length, width and inflow rate using a trial-and-error approach. SIRMOD (Walker, 1998, Walker, 2003) is a full hydrodynamic model applying a volume balance model to determine infiltration characteristics for the evaluation, simulation and optimization of irrigation parameters such as cross section, inflow rate and field slope. SISCO (Gillies and Smith, 2015, Smith *et al.*, 2009) applies a hydrodynamic solution to the Saint-Venant equation for the simulation, evaluation, calibration and optimization of surface irrigation and accounts for temporal variations in inflow rates and spatial variability in soil infiltration, surface roughness, slope and geometry (Gillies and Smith, 2015). WinSRFR (Bautista *et al.*, 2009b, Bautista *et al.*, 2009a) is a hydraulic simulation software for the improvement of design and operation through optimization of field dimensions and finding the best combination of inflow rate and cut-off time. Although the WinSRFR and SIRMOD software are incapable of performing optimization of the performance indicators, they are the most widely-used and comprehensive models (Adamala *et al.*, 2014). WinSRFR has the additional advantage of being computationally faster and freely available (Koech *et al.*, 2010). This research applies the WinSRFR model for the evaluation of field design and operation in the study area.

The research utilized primary data for soil texture, soil moisture before and after irrigation, soil water deficit, and field water capacity. It also used available secondary data collected in 2015 for inflow measurements, field geometry of a pilot field in Kassala in GAS (HRC, 2016), in addition to the observed advance and recession times by the GAS water master. Major difficulties were encountered in obtaining precise field measurements for field infiltration properties due to the inaccessibility of flooded fields (Figure 4.4).

Figure 4.4: An irrigation field during flood irrigation.

In the 1970s, the National Resources and Conservation Services (NRCS) of the US department of Agriculture developed the concept of intake families (NRCS infiltration families), as a way of categorizing infiltration behaviour for similar soils (USDA (1974). The observed soil type of the pilot field was silty clay and can be described by NRCS intake family 0.25. These data were necessary for flow evaluation, field design and simulation of alternative strategies using WinSRFR 4.3.1.

The model combines unsteady flow simulation, evaluation, parameter estimation, system design and optimization of operation (Bautista *et al.*, 2010) using four main windows: event analysis, design analysis, operation analysis and simulation. An overview of the technical elements of the model is found in Bautista *et al.* (2009b). The event analysis provides an assessment of the observed irrigation event, while design and operation alternatives are provided by the design and operation windows, respectively. The simulation window examines different scenarios and conducts a sensitivity analysis (Bautista *et al.*, 2009a).

An important input in WinSRFR is the required irrigation depth (D_{req}) which influences the performance estimates (Bautista *et al.*, 2006). The Required Depth (net irrigation depth) is the depth of water needed to replace the root zone soil water deficit (Bautista *et al.*, 2012), and can be calculated using the following formula (e.q. 4.1) (Salazar *et al.*, 1994):

$$D_{req} = ET_C - P_e - GW - D_b \qquad\qquad \text{e.q. 4.1}$$

Where: ET_C is the crop evapotranspiration (crop water requirement) defined as the depth of water needed to meet the water loss through evapotranspiration (mm), P_e is the effective rainfall (mm), GW is the groundwater contribution through capillary rise (mm), and D_b is the available stored water before irrigation (mm). This study ignored the GW contribution as no study has been done to quantify this parameter, although it has been claimed that GW contributes to satisfy crop needs through capillary rise due to fine soil texture with high silt content and moderate permeability (Khalid, 2009). GW contributes to the CWR during the crop development stage especially in September when the moisture in the upper layer is depleted. As the type of soils in spate flood plains allows replenishment of moisture through upward flux (capillary rise) from the deep soil to the upper dried boundary of the soil profile, crop water stress is thus to some extent avoided.

The measured available soil moisture before irrigation D_b showed a dry soil sample. Avelino (2012) studied the crop water requirement (CWR) for sorghum in the Kassala Block using CROPWAT and found the CWR amounted to 519 mm without an effective rainfall contribution and 400 mm with a rainfall contribution. Steenbergen *et al.* (2010) reported a value of 500 mm of stored soil moisture depth from a single irrigation in GAS. Myers (1980) studied the root system of a sorghum grain crop and found that 78% of the root length were in the 40 cm soil depth. Therefore $D_{req} = 600$ mm was selected for the analysis.

Since WinSRFR can perform only 1-D flow analysis, the study assumed a uniform slope which could vary from the actual condition due to poor land levelling that results in surface irregularities and cross slopes. A 2-D dimensional simulation studies could better simulate the effect of variation in bed level, minimum possible inflow rate, and minimum required upstream depth (Playán *et al.*, 1994). Therefore, the assumptions made are uniform flow, graded field, no cross slope, uniform inflow rate that is uniformly distributed over width, and homogeneous soil surface roughness. The simulations resulting from this study do not account for the actual practice of manually spreading the water from upstream towards the downstream end field and the use of field spurs which helps in water distribution from lower to higher level parts. The performance of an irrigation event was evaluated using efficiency, adequacy, and uniformity in terms of application efficiency (AE), irrigation adequacy (AD), and distribution uniformity (DU), respectively. Flow advancement along the field was also used in the analysis.

AE is the fraction of the total volume of water delivered to the farm or field to that which is stored in the root zone to meet the crop evapotranspiration needs (Irmak *et al.*, 2011). AE can also be described in terms of water depth (e.q. 4.2). (FAO, 1988) reported field application efficiency for irrigation schemes using surface irrigation as follows: 60-50% as good, 40% as reasonable, and <30% as poor. AE is considered the primary criterion in the design and management of border irrigation systems (Zerihun et al., 2005). It gives information of water losses through deep percolation and runoff. Although AE has been widely used as a decision criteria, the indicator has been criticized as being subjective to the user choice for the required depth (Anwar et al., 2016), and the possibility of achieving high AE, yet inadequate irrigation performance by applying less than CWR to minimize losses to deep percolation (Irmak et al., 2011).

$$AE = \frac{\text{Average depth of infiltrated water stored in the root zone}}{\text{Depth of water applied to the field}} = \frac{D_z}{D_{app}} \qquad \text{e.q 4.2}$$

AD provides an estimate of irrigation adequacy (under-irrigation, proper irrigation, over-irrigation) to deliver the amount of water required to adequately irrigate crops (Burt et al., 1997). In this research AD represents the ratio (or percentage) of the least-quarter average depth to the desired target depth (e.q. 4.3). Least-quarter depth (D_{lq}) is defined as the average depth for quarter of field receiving the least infiltrated depth. D_{lq} has been used successfully in irrigated agriculture (Burt et al., 1997). A field will be under-irrigated for AD<1, properly irrigated for AD=1, and over-irrigated for AD>1 (Burt et al., 1997).

$$AD = \frac{\text{Least−quarter average infiltrated depth}}{\text{Water required in the root zone}} = \frac{D_{lq}}{D_{req}} \qquad \text{e.q. 4.3}$$

The DU of the low quarter (ratio or percentage) is defined as the average depth infiltrated in the least-quarter of the field divided by the average depth infiltrated over the entire field (Irmak et al., 2011) (e.q. 4.4). DU can indirectly affect the irrigation performance with non-uniform water application which results in crops being water-stressed or oxygen-stressed, and hence there is a reduction in crop yield (Irmak et al., 2011). Flow advancement represents the ratio of length of flow advance to the total field length.

$$DU = \frac{\text{Least−quarter average infiltrated depth}}{\text{Average infiltrated depth}} = \frac{D_{lq}}{D_{inf}} \qquad \text{e.q. 4.4}$$

Bautista et al. (2009b) recommended that a sensitivity analysis be conducted with unsteady flow simulation to assess the robustness of solutions due to variations in the design parameters, infiltration and roughness characteristics. Hence, the sensitivity of the performance of the selected field layout at different inflow rates and application time was measured under different uncertain parameters.

Analysis of the scenarios

The objective of the scenario analysis was to optimize the practical field layout configuration and operation practice, as one method for reducing flood risk due to flow variability in GAS. The measured data used for the evaluation and calibration analysis were: field dimensions (8,400×500 m), daily inflow hydrographs, observed advance and recession curves, field slope (13.8 cm/km), observed average inflow rate (2,270 l/s), cut-off time T_{co} (600 hrs), soil type (silty clay) and estimated Manning roughness (n= 0.08), which is within the basic recommended range of the field roughness conditions 0.06 to 0.09 (Bautista *et al.*, 2009a, Salahou *et al.*, 2018, Zhang *et al.*, 2006).

Three strategies were investigated: 1- time management, 2- field design and time management, and 3-field design and flow management. Figure 4.5 illustrates the studied strategies while Table 4-2 lists the potential cut-off times T_{co} and inflow rates using very high, high, average, low and very low inflow rates that are assumed to occur during large, medium and low flood seasons. The design flow of the irrigation fields in GAS based on experience is such that 1,000 l/s is sufficient to irrigate 210 ha on heavy soils and 100 ha on light soil for a duration of 25-28 days. A detailed description of the strategies is presented as follows:

Time management strategy: Under this strategy, the impact of change of T_{co} while maintaining the current field design (dimensions) was examined under variable inflow rates representing extreme and normal conditions such as large, medium and low floods. Performance measures were obtained for different scenarios of potential inflow rates (2,600, 2,400, 2,100, 1,600, and 1,000 l/s) and potential cut-off times T_{co} (672, 600, 432, 300, 240 and 168 hours) representing extended and short flood possibilities.

Field design and time management strategy: This strategy illustrates improvements in the current field design through field division into two equal areas; either vertically divided (8,400×250 m), or horizontally divided (4,200×500 m). Total potential inflow rates were applied in sequence for each sub-divided field. Therefore the total planned T_{co} was divided equally between the two subfields. Performance indicators for the irrigation scenarios were obtained under potential inflow rates (2,600, 2,400, 2,100, 1,600, and 1,000 l/s) and 0.5 T_{co} (336, 300, 240, and 168 hrs).

Field design and inflow management strategy: In this strategy, the potential inflow rates were divided equally between sub-fields. i.e. two vertically divided fields (8,400×250 m) or two horizontally divided fields (4,200×500 m). Performance indicators were generated for different scenarios of 0.5 Q and T_{co}. Irrigation of the two sub-fields started and ended at exactly the same time to maintain the total planned T_{co}. Possible inflow rates entering each subfield were 1,300, 1,200, 1,100, 800, 500 l/s, while possible T_{co} were 672, 600, 432, 300, 240, and 168 hrs.

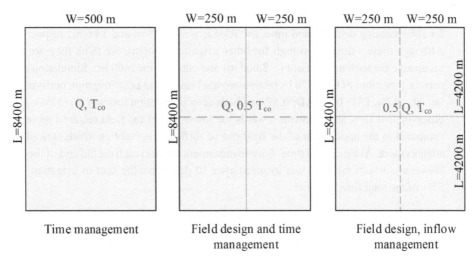

Figure 4.5: Study variables.

Table 4-2: Tested variables

Flood size	Time management		Field design & time management		Field design and flow management	
	Flow rate (l/s)	T_{co} (hrs)	Flow rate (l/s)	T_{co} (hrs)	Flow rate (l/s)	T_{co} (hrs)
Very high	2,600	672, 600,	2,600	336, 300,	2,600/2	672, 600,
High	2,400	480, 432,	2,400	240, 168	2,400/2	480, 432,
Average	2,100	336, 300,	2,100		2,100/2	336, 300,
Low	1,600	240, 168	1,600		1,600/2	240, 168
Very low	1,000		1,000		1,000/2	

Note: m denotes meter, T_{co} denotes cut-off or application time.

4.3 RESULTS AND DISCUSSION

4.3.1 Model calibration

The selection of the infiltration parameters for the event analysis was obtained after a lot of trial and error for several empirical infiltration functions presented by the WinSRFR model. This was done through comparison of the advance and recession curves with those produced by the model. The judgment was based on prior knowledge of the flow behaviour and observations. The results of the calibration of the observed advance and observed recession curves using the NRCS infiltration family 0.25 showed acceptable agreement with the simulated values, as illustrated in Figure 4.6. The root mean square error (RMSE) for advance distance and time were 32.7 m and 30.4 hr, respectively, while for the recession distance and time, the RMSE were 41.8 m and 14.6 hr, respectively. Although these values seem high for other irrigation systems, we think they are quite acceptable for such a long field (> 8,000 m) and cut-off time (600 hr). Simulation of the current conditions of Q=2,270 l/s (above average) resulted in poor irrigation performance, i.e. AE =47%, DU= 0.45, AD=0.87, and high deep percolation losses (DP) (53%). It was observed that flow advancement towards the lower end of the field required more time compared to the upper parts of the field due to surface irregularities which slowed flow advancement. At the cut-off time, flow advancement did not reach the tail end of the field. However, 70% of the field was irrigated after 10 days from the start of irrigation using 55% of the total flow volume.

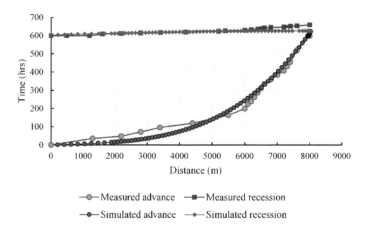

Figure 4.6: Observed and simulated advance and recession curves using the NRCS intake family 0.25.

4.3.2 Performance of improved strategies

a- Time management strategy:

A hydraulic simulation of the WinSRFR model for the time management strategy provided the performance of possible improvements on the current field through change in the application time (T_{co}). The results as shown in Table 4-3 indicated that fair DU (≥0.7) could only be obtained when high inflow rates applied at high T_{co} (≥600 hrs). Similarly, full flow advance only occurred at high T_{co} which is also associated with over-irrigation conditions (AD > 1). DP losses were encountered at locations near the intake and reduced towards the tail-end locations. This is attributed to the upper-lower irrigation approach used to distribute irrigation water by gravity which exposes the top head parts of the fields to over-irrigation, sediment accumulation and large DP losses. Nevertheless,

González-Cebollada *et al.* (2011) attributed this phenomenon to the impact of surface topography. Therefore, increasing the application time to relatively improve performance will not be practical under existing water scarcity and demand pressure. The optimal strategy is the strategy which combined the best possible performance of all indicators. This strategy showed poor optimal performance of all the indicators under large, medium and small flood seasons. The optimal management strategies during large flood seasons were observed at T_{co}=480 hrs for Q=2,600 l/s, and at T_{co}=600 hrs for Q=2,400 l/s, while medium and small flood seasons showed poor performance at all the application times. In general, the optimum performance showed over irrigation and poor DU during large floods, and under irrigation and poor DU during medium and low floods. Zero values indicate the absence of water at the least-quarter depths. However, based on field observations of average flow conditions, farmers could irrigate 70% of the field using 50% of the flow volume for a period of 10 days, and save more irrigation time and flows to irrigate downstream fields. This strategy can significantly reduce deep percolation losses in the system.

Table 4-3: Performance of the time management strategy

Strategy	Variables			Irrigation performance			
	Field layout (m×m)	Inflow rate (l/s)	T_{co} (hrs)	Application efficiency (%)	Distribution uniformity (ratio)	Adequacy (ratio)	Flow advance (ratio)
Time management	8,400×500	2,600	672	40	0.76	1.82	1.0
			600	45	0.72	1.58	1.0
			480	**55**	**0.62**	**1.1**	**1.0**
			432	59	0.55	0.87	0.99
			336	69	0.35	0.44	0.93
			300	74	0.27	0.3	0.89
			240	84	0.13	0.11	0.83
			168	99	0.00	0.00	0.73
		2,400	672	43	0.65	1.49	1.0
			600	**47**	**0.59**	**1.21**	**1.0**
			480	55	0.42	0.7	0.94
			432	59	0.37	0.51	0.92
			336	69	0.17	0.2	0.86
			300	74	0.11	0.11	0.81
			240	84	0.01	0.01	0.77
			168	99	0.00	0.00	0.66
		2,100	**672**	**44**	**0.34**	**0.68**	**0.92**
			600	48	0.26	0.47	0.88
			480	55	0.12	0.17	0.82
			432	59	0.07	0.08	0.81
			336	69	0.00	0	0.75

	300	74	0.00	0.00	0.71
	240	84	0.00	0.00	0.68
	168	99	0.00	0.00	0.61
1,600	**672**	**44**	**0.00**	**0.00**	**0.69**
	600	47	0.00	0.00	0.68
	480	55	0.00	0.00	0.63
	432	58	0.00	0.00	0.62
	336	69	0.00	0.00	0.57
	300	74	0.00	0.00	0.54
	240	84	0.00	0.00	0.51
	168	99	0.00	0.00	0.45
1,000	**672**	**44**	**0.00**	**0.00**	**0.44**
	600	47	0.00	0.00	0.42
	480	55	0.00	0.00	0.40
	432	58	0.00	0.00	0.38
	336	69	0.00	0.00	0.36
	300	74	0.00	0.00	0.33
	240	84	0.00	0.00	0.32
	168	99	0.00	0.00	0.29

Note: m denotes meter, T_{co} denotes cut-off or application time.

b- Field design and time management strategy:

Simulation results of improved field design and time management (Q, 0.5 T_{co}) using vertical division (8,400×250 m) and horizontal division (4,200×500 m) resulted in a similar performance during large, medium, and small flood seasons (Figure 4.7). The analysis proceeded with an investigation of a vertically divided field (8,400×500 m) due to several practical reasons which eliminate the horizontal division choice. For example, in the horizontal division, there is a need to implement a new field canal that extends for more than 4 km with all the risks of conveyance and seepage losses, frequent maintenance requirements, and the high risk of conflicts for farmers due to rule breaking. Additionally, extensions of conveyance canals that are too long result in large deep percolation losses in the permeable alluvial of the unlined canals found in spate irrigation systems (Tesfai and Stroosnijder, 2001). The simulation of the selected vertical division revealed that during the above average flows, DP and RO (Runoff) losses accounted for almost 50% of the applied water at T_{co} >240 hrs, while RO losses diminish at smaller T_{co}. Similarly, reduction in flow rates to below average results in losses attributed solely to DP. Therefore, an increase in AE was obtained at reduced T_{co} and inflow rates. Since minimizing DP losses may result in irrigation depth less than required to satisfy crop needs (Irmak et al., 2011), a trade-off needs to be made for the possible accepted values. DU gave good results for almost all the flow rates except at very low flows of Q≤ 1,000 l/s. The optimum performance of all indicators were found at T_{co}=168 hrs for very high and high flows, T_{co}=240 hrs for average flows, and T_{co}= 300 hrs for low and very low flows. Table 4-4 lists the performance of the field design and time management strategy.

Table 4-4: Performance of the field design and time management strategy

Strategy	Variables			Irrigation performance			
	Field layout (m×m)	Inflow rate (l/s)	T_{co} (hrs)	Application efficiency (%)	Distribution uniformity (ratio)	Adequacy (ratio)	Flow advance (ratio)
Design & time management	8,400×250		336	40	0.96	1.80	1.00
		2,600	300	45	0.95	1.47	1.00
			240	56	0.94	1.22	1.00
			168	**78**	**0.91**	**0.89**	**1.00**
		2,400	336	43	0.95	1.57	1.00
			300	49	0.94	1.43	1.00
			240	61	0.92	1.16	1.00
			168	**83**	**0.88**	**0.86**	**1.00**
		2,100	336	50	0.92	1.48	1.00
			300	56	0.90	1.32	1.00
			240	**70**	**0.87**	**1.08**	**1.00**
			168	92	0.81	0.76	1.00
		1,600	336	65	0.76	1.12	1.00
			300	**71**	**0.72**	**0.96**	**1.00**
			240	84	0.62	0.67	1.00
			168	99	0.35	0.26	0.92
		1,000	336	70	0.00	0.00	0.71
			300	**74**	**0.00**	**0.00**	**0.68**
			240	84	0.00	0.00	0.64
			168	99	0.00	0.00	0.57

Note: m denotes meter, T_{co} denotes cut-off or application time.

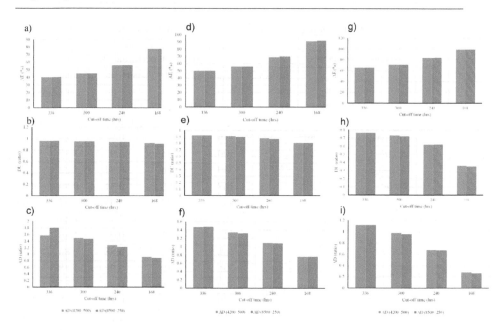

Figure 4.7: Performance indicators of vertical and horizontal field division during: a large flood season (a, b, c), a medium flood season (d, e, f), and a small flood season (g, h, i).

c- Field design and flow management strategy:

Simulation results of 0.5Q and T_{co} of the vertically (8,400×500 m) and horizontally (4,200×500 m) divided fields resulted in similar performance during large, and medium flood seasons. However both options resulted in poor performance in small flood seasons. Therefore, the analysis proceeded with an investigation of a vertically divided field. Table 4-5 lists the performance of field design and flow management strategy. The results of this strategy were found to be comparable to the time management strategy. The optimal performance indicators were observed at T_{co} =480 hrs for Q=1,300 l/s (very high flow), T_{co}=600 hrs for Q=1,100 l/s (high flow), and T_{co}=672 for Q≤ 1,100 l/s (average and low flow conditions). Poor DU and over-irrigation results were observed at the optimum condition. Higher AE is found to be associated with poor irrigation performance of AD and DU, particularly at small T_{co}.

Table 4-5: Performance of the field design and flow management strategy

Strategy	Variables			Irrigation performance			
	Field layout (m×m)	Inflow rate (l/s)	T$_{co}$ (hrs)	Application Efficiency (%)	Distribution uniformity (ratio)	Adequacy (ratio)	Flow advance (ratio)
Design & flow managem ent	8,400× 250	1,300	672	40	0.76	1.82	1.00
			600	45	0.72	1.58	1.00
			480	**55**	**0.62**	**1.1**	**1.00**
			432	59	0.55	0.87	0.99
			336	70	0.35	0.44	0.92
			300	74	0.27	0.31	0.78
			240	84	0.13	0.11	0.77
			168	99	0	0	0.74
		1,200	672	43	0.65	1.49	1.00
			600	**47**	**0.59**	**1.21**	**1.00**
			480	55	0.42	0.7	0.95
			432	59	0.34	0.51	0.93
			336	69	0.17	0.2	0.86
			300	74	0.11	0.11	0.81
			240	84	0.01	0.01	0.76
			168	99	0	0	0.68
		1,100	**672**	**44**	**0.45**	**0.96**	**0.95**
			600	48	0.37	0.7	0.92
			480	55	0.21	0.32	0.87
			432	59	0.15	0.2	0.85
			336	69	0.03	0.03	0.79
			300	74	0	0	0.73
			240	84	0	0	0.69
			168	99	0	0	0.63
		800	**672**	**44**	**0**	**0**	**0.69**
			600	47	0	0	0.68
			480	55	0	0	0.63
			432	58	0	0	0.62
			336	69	0	0	0.57
			300	74	0	0	0.50
			240	84	0	0	0.49
			168	99	0	0	0.45
		500	**672**	**44**	**0**	**0**	**0.44**
			600	47	0	0	0.42
			480	55	0	0	0.39
			432	58	0	0	0.39
			336	69	0	0	0.36
			300	74	0	0	0.50
			240	84	0	0	0.49
			168	99	0	0	0.45

Note: m denotes meter, T$_{co}$ denotes cut-off or application time.

4.3.3 Optimum choice

The field design and time management strategy resulted in higher performance indicator values compared to the other strategies. Table 4-6 lists the performance results for the optimum conditions from all the strategies. The selection of a feasible yet satisfactory strategy resulted in different application times T_{co} based on the flood size received, which confirms the flexibility in operation when dealing with spate irrigation systems. Similar findings were obtained by (Taghizadeh *et al.*, 2013) who found that AE could be increased by managing inflow rate Q and T_{co}. The optimum performance during very high-high (2,600-2400 l/s), average (2,100 l/s) and low (1,600 l/s) inflow rates were obtained at T_{co} =168, 240 and 300 hrs, respectively. Figure 4.8 compares the optimum performance of the time management strategy and field design & time management strategy under the best combination of Q and T_{co}. It was evident that the AD of irrigation, DU, and AE were substantially improved in the field design and time management strategy compared to the other strategies. In addition, the three performance indicators showed good, sustained and uniform performance for a wide range of inflow rates (2,600-1,600 l/s) unlike the time management strategy which failed to achieve an acceptable combination of all the performance indicators.

It is worth noting that the field design and time management strategy requires consideration of the RO losses during high and average flows through providing drainage paths directed for reuse. Increasing the AE means less water is provided to the crop and groundwater storage while it does not guarantee improvement of irrigation adequacy or uniformity efficiency (Irmak *et al.*, 2011). This was also confirmed by Burt *et al.* (1997) who demonstrated that water availability could only be improved by decreasing water consumption.

This research showed that field size and application/cut-off time T_{co} were the main elements which had substantial influence on irrigation performance unlike the findings by (Gillies *et al.*, 2010) who recommended higher inflow rates as an operation strategy to improve surface irrigation performance in furrow systems.

Table 4-6: Optimum application time at different flow rates

Strategies	Variables			Irrigation performance			
	Field layout (m×m)	Inflow rate (l/s)	T_{co} (hrs)	Application efficiency (%)	Distribution uniformity (ratio)	Adequacy (ratio)	Flow advance (ratio)
Time management	8,400×500	2,600	480	55	0.62	1.10	1.00
		2,400	600	47	0.59	1.21	1.00
		2,100	672	44	0.34	0.68	0.92
		1,600	672	44	0.00	0.00	0.69
		1,000	672	44	0.00	0.00	0.44
Design & time management	8,400×250	2,600	168	78	0.91	0.89	1.00
		2,400	168	83	0.88	0.86	1.00
		2,100	240	70	0.87	1.08	1.00
		1,600	300	71	0.72	0.96	1.00
		1,000	300	74	0.00	0.00	0.68
Design & flow management	8,400×250	1,300	480	55	0.62	1.10	1.00
		1,200	600	47	0.59	1.21	1.00
		1,100	672	44	0.45	0.96	0.95
		800	672	44	0.00	0.00	0.69
		500	672	44	0.00	0.00	0.44

Note: m denotes meter, T_{co} denotes cut-off or application time.

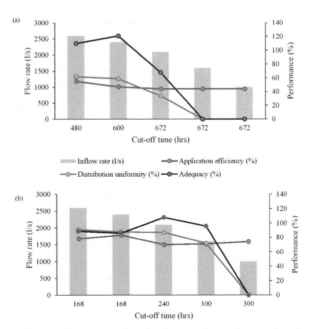

Figure 4.8: Irrigation performance of optimal operation using a) the time management strategy, b) the field design & time management strategy.

4.3.4 Sensitivity analysis

A sensitivity analysis of the proposed operation strategy was conducted to test the robustness of the strategy under possible changes in the inputs and design parameters. We investigated the impact of changes in the slope (S), Manning's roughness (n), and required depth (D_{req}) on the performance. Tested variables for different parameters were as follows: slope= 0.001, 0.0015, 0.002, 0.0025, 0.003; Manning roughness coefficient= 0.04, 0.08, 0.12, 0.16, 0.2; required depth= 400, 500, 600, 700, 800, 900, and 1,000 mm. The analysis indicated that changes in the values of n showed no significant changes in the performance values. The results were in line with the findings of Nie *et al.* (2014) who recommended the safe use of average field roughness as a representative value for the whole field. Similar findings were obtained for changes in slope values under different inflow rates. González *et al.* (2011) showed that at very low inflows the distribution uniformity was largely insensitive to the slope. On the other hand, the sensitivity analysis of D_{req} during small, medium, and large flood seasons is illustrated in Figure 4.9 (a-c). During large floods, D_{req} showed significant change in the adequacy performance and resulted in over irrigation at D_{req}< 500 mm and significant under irrigation at D_{req}> 800 mm. For medium and small flood seasons, over irrigation occurred at D_{req}< 600 mm, while under irrigation occurred at D_{req}> 900 mm. Since D_{req} is not a defining parameter for DU, stable values for DU were observed. Therefore, any change in the D_{req} due to change in the cropping pattern or other factors needs the establishment of new operation strategies. This was confirmed by (Saher *et al.*, 2014) who indicated that traditional water rights should be modified with a new cropping pattern.

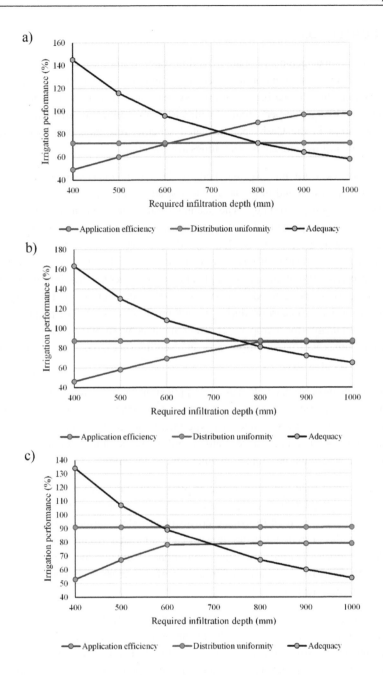

Figure 4.9: Sensitivity to irrigation depth (D_{req}) of the selected operational strategy during: a) small flood, b) medium flood, and c) large flood seasons.

4.4 CONCLUSION

Operation and management of spate irrigation water, using flash floods from ephemeral rivers, is becoming more important to farming communities whose livelihood conditions are worsening by climate-based farming systems and increased food insecurity. To maintain good irrigation performance with sustainable utilization of limited and scarce water resources in the Gash spate-irrigated scheme, recommendations on ranges of possible field inflows and application times are essential.

Three strategies were investigated, namely 1) a time management strategy; 2) an improved field design and time management strategy, and 3) an improved field design and flow management strategy. The second strategy resulted in higher performance indicator values compared to the other strategies. Using improved field design and flow management strategies, this research achieved high performance indicators compared to current field design under different application times and inflow rates. The proposed strategy can save 40% of the current application time, under the high inflow rate scenario (occurring during a large flood season), and a 20% reduction in time under the average inflow rate scenario (occurring during a medium flood season). The proposed strategy could be sustainable for a crop water requirement between 500 to 800 mm. A list of application times at different inflow rates representing very large, large, average and low flow rates were provided.

This research showed that field size and application time T_{co} were the main elements which had substantial influence on irrigation performance. The modelling outcomes confirmed that the farmers' indigenous experiment, though without a scientific study, on vertical division of a large-sized field is indeed successful in improving irrigation performance, and could be adopted in other similar conditions.

Under the current field condition, reducing 30% of the field area at the tail end can save 60% of the application time under average inflow rates and can reduce 50% of water losses which could be used to irrigate downstream fields. Further irrigation of tail-end fields using the current upstream downstream approach resulted in increased deep percolation losses. During low flow conditions, reducing the command area can significantly increase the chance of irrigation of the upper-head field located further downstream the irrigation offtake.

The uncertainty involved in field measurements and infiltration parameters that are estimated by the model could be reduced through further research and extensive field measurements for infiltration characteristics of the study area. Additionally, lack of data on GW contribution in the enhancement of irrigation performance through capillary rise, the limitation of WinSRFR to model the actual non-steady flow and field conditions, and the complex flow behaviour around the field spurs add to the limitations of this study.

5

FLEXIBILITY AS A STRATEGY TO COPE WITH UNCERTAIN WATER SUPPLY IN SPATE IRRIGATION[6]

Unpredictable flash floods in ephemeral rivers are the water source for spate irrigation systems. An important element for the success and sustainability of spate irrigation systems is their ability to cope with highly uncertain water supply and high sediment load. Flexibility is considered as one of the key ingredients of coping strategies. However, the concept of flexibility in the context of spate irrigation systems is poorly defined. A framework to assess and operationalize flexibility in spate irrigation is lacking. In this paper we develop a conceptual framework through answering four principle questions and exploring eight flexibility characteristic features and five sub-features. The flexibility of traditional, improved and modernize spate irrigation systems to cope with high, low and untimely flood events is explored. Flexible spate irrigation systems are highly dependent on system capabilities to deal with uncertainty and enable adjustments to the change. The framework can be used as a guideline to water managers, farmers and decision makers for assessing and providing flexibility in the spate irrigation systems.

[6] This chapter is based on a paper to be submitted to the Journal of Arid Environment: FADUL, E., DE FRAITURE & C., MASIH, I. Flexibility as strategy to cope with uncertain water supply in spate irrigation systems.

5.1 INTRODUCTION

Spate irrigation is a type of flood-based farming that makes use of highly variable and seasonal flash floods in ephemeral rivers, often one of the few available water sources in arid and semi- arid regions. Spate irrigation systems need to cope with a high level of uncertainty and unpredictability regarding flood size and timing. Additionally, spate irrigation systems are exposed to continuously changing socioeconomic and physical conditions. Changes in physical conditions include alteration in river (or wadi) morphology due to erosion and sediment deposition, increase in command area level and the destruction of irrigation infrastructure. The socio economic dynamics include changes in policies that affect agricultural production, access and distribution of water, access to technology and seasonal migration.

A big challenge in the planning of spate irrigation systems is the design of engineering works. Because of lack of historical record of river flow and sediment transport data, most spate irrigation systems are designed based on empirical methods and assumptions. Even if historical records are available, the design of hydraulic structures using deterministic and probabilistic forecasts assumes hydrologic stationarity (Major and Frederick, 1997, Schulz *et al.*, 2000) which may not be valid in areas with very high climate variability and change (Frederick *et al.*, 1997, Lempert, 2003). Seasonal flash floods, the main water source in spate irrigation, are characterized by sequences of short duration high flood peaks, high flow velocity, and large sediment load that could lead to damage to life, property and infrastructure (Borga *et al.*, 2011, Creutin *et al.*, 2013). At other times, these floods may reduce to small flows which are unusable for irrigation. With these high level of changeability and unpredictability in water supply, it is challenging to plan, operate and take actions. Farmers and water managers are uncertain about pre-season decisions on irrigation plans, irrigable area, irrigation duration, decision on operation of intakes and offtakes, level of maintenance, and level of investment on farming activities and land preparation.

To cope with uncertainties in water resources and flood water management, several studies recommend maximizing flexibility in design and operation (IPCC, 2007, Turral *et al.*, 2011, Huang *et al.*, 2010). For example, the FAO publication "Guidelines for spate irrigation" recommended a flexible approach to spate irrigation improvement to cope with changing conditions, frequency and severity of extreme events and uncertainty (Steenbergen *et al.*, 2011). Flexibility of spate irrigation systems (traditional, improved, modernized) depend on the availability of viable actions and decisions at the river diversion, the canal network and irrigated fields. So-called traditional spate systems were generally regarded as the most flexible, having flexible water rights (Van Steenbergen, 1997, Kamran and Shivakoti, 2013), flexible irrigation turns (Haile *et al.*, 2011, Haile *et al.*, 2005c), and flexible control of intakes (Steenbergen *et al.*, 2011). For example, Haile *et al.* (2005c) described flexible water rights and rules between different farmers in Wadi Leba in Eretria such as flexible land & water scheduling, proportional distribution of flood water, flood water distribution based on flood size; and rules on canal breaching to avoid downstream damage (Haile A., 2003). Van Steenbergen (1997) recommended the use of flexibility in the engineering design of spate irrigation systems to accommodate water supply variability. The so-called modernized systems are considered the most rigid, with claims that the lack of flexibility contributed to their failure (Haile *et al.*, 2003, Van Steenbergen, 1997, Haile *et al.*, 2011, Steenbergen *et al.*, 2011, Haile *et al.*, 2005a).

While flexibility is widely recommended as strategy to cope with unpredictable water supply in spate irrigation systems, there is no in-depth analysis on the subject. A systematic framework to better understand flexibility and to provide a conceptual approach to formulate flexible real options in the planning, design and operation of spate irrigation systems is lacking. Therefore, in this paper we assess the underlying assumptions regarding flexibility of different types of spate irrigation technologies. First, we develop a conceptual framework for inclusion of flexibility in the context of spate irrigation. Then, we assess flexibility of traditional, improved traditional and modernized spate irrigation systems. Lastly we analyse the different flexibility options to cope with unpredictable, uncertain and highly variable water supply.

The research was based on data collected from primary and secondary sources. This included more than 100 interviews with farmers, water user associations (WUA) and water managers from the Gash spate irrigation system (GAS) in Sudan (Fadul et al., 2019); literature review on existing technologies in different countries such as Yemen (SIN, 2013b, Haile *et al.*, 2011), Ethiopia (Libsekal *et al.*, 2015, Castelli *et al.*, 2018, Tadesse and Dinka, 2018), Pakistan (Haile *et al.*, 2011, Khan *et al.*, 2014), Eretria (Haile *et al.*, 2011), Iran (Kowsar, 2011b), Morocco (Oudra, 2011), Myanmar (SIN, 2013a), Sudan (Van Steenbergen *et al.*, 2011, Steenbergen *et al.*, 2011, Zenebe *et al.*, 2015b); and information gathered during a regional workshop on farmer to farmer knowledge sharing conducted in 2012 in Sudan (Fadul *et al.*, 2012).

5.1.1 Technology options in spate irrigation system

Different technologies are being used for river diversion, irrigation conveyance network, and field application. Usually they are categorized as traditional, improved traditional, and modernized systems (Table 5-1) (Van Steenbergen *et al.*, 2010, Haile *et al.*, 2011, Komakech *et al.*, 2011, Kowsar, 2011a, Steenbergen *et al.*, 2011, Haile *et al.*, 2005b, Haile *et al.*, 2005a, Van Steenbergen, 1997).

Traditional spate system

The practice of spate irrigation has a long history. Traditional systems are built and maintained using farmers' indigenous knowledge and skills, local materials and resources without external support (Haile *et al.*, 2011, Van Steenbergen *et al.*, 2010). Traditional intakes and diversion weirs are designed to minimize interference with the flow path of floods in river channel. The weirs made of sticks and stones are washed away during medium to large floods and hence prevent large and potentially destructive, high sediment-laden floods from entering the canal system (Van Steenbergen *et al.*, 2010). The diversion system consists of a series of small diversion weirs which allow downstream users to divert flood water even when large floods breached upstream weirs. Traditional intakes need frequent maintenance and reconstruction after each large flood, which requires strong collaboration between all farmers (Camacho, 1987, Steenbergen *et al.*, 2011). Additionally, the practice of field to field irrigation, whereby flood water has to pass through the same top field to irrigate the next fields, results in over-irrigation and sediment built-up in head fields (Steenbergen *et al.*, 2011).

Improved traditional systems

To overcome these drawbacks, in some traditional systems improvements have been made to ensure less labour intensive and relatively permanent structures, without major alterations in the existing practices (Tadesse and Dinka, 2018, Van Steenbergen *et al.*, 2010, Castelli and Bresci, 2017, Castelli *et al.*, 2018, Haile *et al.*, 2005a, Haile *et al.*, 2005b). These improvements include the use of adjustable stone or masonry weirs with breach or overflow sections; improved diversion bunds with bed stabilizers to replace the earthen and stone structures; reinforced intakes with bricks and mortar; and structures to better control water distribution along the main canal. At field level an improved field to field system is introduced through the use of drop of structures made of masonry to reduce scour risk.

Modernized systems

Modernized system are characterized by the use of permanent structures made of concrete, a sediment exclusion system and an irrigation network consisting of one single main canal and secondary, tertiary and field canals which allow supplying water to individual fields. Modernized system are sufficiently robust to tolerate variability and divert both advance and recession flow including (reasonably) high floods which cannot be diverted by traditional systems. Nevertheless, if the existing rules and practices are not acknowledged during design and operation, modernized system will lead to unfair water distribution by favoring the upstream farmers with more water access (Haile *et al.*, 2011). Failure of modernized systems in countries such as Ethiopia, Yemen and Pakistan, occurred partly due to the diversion of large floods with high sediment load, poor design, and the low involvement of farmers in the development, design and construction process (Castelli and Bresci, 2017, Libsekal *et al.*, 2015, Van Steenbergen *et al.*, 2010, Van Steenbergen, 1997, Haile *et al.*, 2011, Oosterbaan, 2010, Komakech *et al.*, 2011). Traditional systems are limited to shallow depth rivers and wadis (Zaqhloel, 1987, Mu'Allem, 1987) while modernized systems can tap from deeper rivers (Haile *et al.*, 2005a).

Table 5-1: Technology choice in spate irrigation systems with examples of application in various countries

Technology choice			
Location	Traditional system	Improved-traditional system	Modernized-system
Diversion	Soil bunds/spurs (Figure 5.1), open intakes (Yemen, Eretria, Pakistan, Ethiopia), earthen deflecting spurs (Iran, Sudan, Morocco, Myanmar, Yemen), brushwood dam with wooden piles (Iran, Myanmar), diversion weir, terraces with weep holes drainage (Iran)	Reinforced soil bund-spur, Gabion diversion and stabilizers (Iran, Pakistan, Ethiopia), Drop-off structure, Bed stabilizers (Pakistan), Brick mortar, Masonry intakes (Figure 5.2) (Sudan, Iran, Ethiopia), Masonry check dams (Iran) (Wooden stop-log intakes, Stone-gabion open intakes (Yemen, Rejection spillway-Fuse plugs (Eretria, Pakistan, Yemen, Ethiopia)	Concrete weir and gated-concrete intake (Figure 5.3), Automatic intakes (Iran), Settling stilling basin (Yemen, Morocco, Myanmar), Sluice gate (Myanmar) Sediment excluders (Morocco, Yemen, Pakistan, Myanmar), rejection spillway (Yemen, Morocco) Sluice-gate (Myanmar)
Distribution	Several short canals, Flow splitters (Pakistan, Yemen), stone wall (Yemen)	Earthen main canal, Tertiary canal, guide wall (Morocco)	Long single main canal, Controlled-cross structures, Secondary canal, (often), Tertiary canal, Conveyance spreader channel (Iran)
Field	Field to field (Yemen, Eretria)	Individual field inlets (Sudan, Pakistan). Gated offtakes, masonry drop-off structure (Iran), Gated-orifice.	Individual field inlets (Pakistan). Gated field offtake, Field canal, Field spurs, Field embankment, Open/closed-end field, Tail drain (Iran)

Note: The table is compiled from various sources: (Haile *et al.*, 2005a, Camacho, 1987, Castelli *et al.*, 2018, Tadesse and Dinka, 2018, Libsekal *et al.*, 2015, Steenbergen *et al.*, 2010, SIN, 2013a, SIN, 2013b, Kowsar, 2011b, Oudra, 2011).

Figure 5.1: Soil bunds in traditional spate system in Sudan-Toker scheme © Eiman Fadul.

Figure 5.2: Brick mortar intake in improved traditional system in Sudan-Gash scheme © Eiman Fadul.

Figure 5.3: Concrete weir and diversion intake in modernized spate systems in Ethiopia (left) © Eiman Fadul and Yemen (right) © Ahmed Al-Siddig.

5.2 CONCEPTUAL FRAMEWORK FOR FLEXIBILITY IN SPATE IRRIGATION SYSTEMS

There is a large body of literature on the concept of flexibility, outlining general principles and operationalizing in a number of different fields (Anvarifar *et al.*, 2016) such as information technology (Dorsch, 2015), aerospace systems (Saleh *et al.*, 2003), urban and infrastructure development (De Neufville and Scholtes, 2011), emergency management (Ward *et al.*, 2015), software system architecture (Schulz *et al.*, 2000) and water supply and waste water systems (Spiller *et al.*, 2015).

The proposed framework to conceptualize flexibility in spate irrigation systems is adapted from established paradigms for water resources management (DiFrancescoTullos (2014), coastal flood defense (Anvarifar *et al.* (2016) and integrated coastal management (Taljaard *et al.* (2011). Anvarifar *et al.* (2016) identified eight features of flexibility, namely: change, uncertainty, goals, capabilities, mode of response, temporal dimensions, and real options and enablers. Similarly, DiFrancescoTullos (2014) identified five flexibility characteristics which are relevant to water resources systems, namely: slack, redundancy, connectivity, coordination, and adjustability. Both fields of flood defense and water resources management are relevant to the spate irrigation systems in terms of exposure to uncertainty of climate parameters and possible changes of the system. The two frameworks complement each other: the flood defense framework conceptualizes the understanding of flexibility in flood management while the water resources management framework highlights actions and decisions to adjust to changing conditions. Therefore, we propose a combination of the two conceptual frameworks with few modifications to assess flexibility and formulate real options in spate irrigation systems. Following Anvarifar *et al.* (2016) the framework with four self-guiding questions. The combined framework is presented in Figure 5.4 and Table 5-2.

1- Why is flexibility needed? (Q1)
2- What is it that flexibility is required for? (Q2)
3- What are the dimensions of flexibility? (Q3)
4- What needs to be changed or adapted? (Q4)

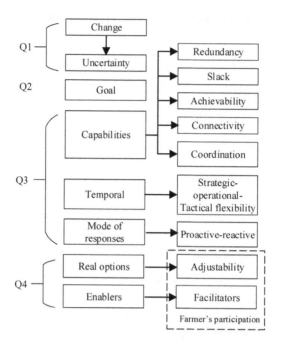

Figure 5.4: Framework for flexibility in spate irrigation systems, adapted from Anvarifar et al. (2016) and DiFrancesco and Tullos 2014.

Table 5-2: Flexibility characteristics in spate irrigation based on principle questions adapted from (Anvarifar *et al.*, 2016)

Questions	Characteristic features	General description
1-Why is flexibility needed?	(a) Change	Internal and external
	(b) Uncertainty	Unpredictable, Unplanned, or Uncertain
2-What is flexibility required for?	(c) Goal	Minimizing water supply risks and maximizing irrigated area
3- What are the dimensions of flexibility?	(d) Systems capability (Slack-Redundancy-Connectivity-Coordination-Achievability)	Range/number of options and quickness of implementation
	(e) Temporal	Strategic, tactical, and operational
	(f) Mode of response	Proactive-Reactive

4-What needs to change or adapted (with farmer involvement)?	(g) Real options (Adjustability)	Actions: Expand, defer, shrink Decisions: Delay, update, add, modify
	(h) Enablers	Sources of alterations in the technical design and management decision

Description of the frame work in the context of spate irrigation

1) Why is flexibility needed?

Change: External sources relate to changes in the system resulting from climatic factors and consequent effects, such as a sudden rise in the peak flood, an extended duration of high flood levels, an unpredicted low flood season, the occurrence of an early or late flood. Types of flood risks in spate irrigation are described in detail by (Fadul *et al.*, 2018, Steenbergen *et al.*, 2011, Haile, 2010). Major <u>internal</u> changes include a sudden breach in the river or canal embankment, escaped intakes, blockage of canals with sediment, failure of diversion weirs and changes in duration of the irrigation season. (Haile, 2007).

Uncertainty results from the <u>unpredictability</u> of flood size, rate, duration and timing. Other sources of uncertainty are related to lack of knowledge and information on technology, maintenance, suitable design approaches that consider the pattern of rising hydrographs (Van Steenbergen, 1997, Steenbergen *et al.*, 2011). Uncertainty might lead to <u>ad hoc</u> operational decisions such as when to open or close an intake while not knowing when the next flood opportunity will come.

2) What is flexibility required for?

Goals represent the desired objectives of flexibility which could handle both upsides and downsides of uncertainty and changes (Anvarifar *et al.*, 2016). They differ by system type and location based on benefits, risks, and costs. For example, in a modernized system in Pakistan, the goals are to increase productivity, improve national food security and livelihood while minimizing failure of irrigation. In the improved system of GAS in eastern Sudan, the goals are to increase irrigated area and minimize failure of structures. In traditional systems the goals often are to minimize risk of diverting high floods. In general, maximizing the opportunities and minimizing the risk in spate irrigation is affected by the type of technology and the level of investment (Table 5-3).

Table 5-3: The need for and goals of flexibility for different types of spate irrigation systems

Flexibility question	Flexibility characteristic features	Traditional system	Improved traditional system	Modernized system
1. Why is flexibility needed?	(1) Change	Water level increase in the river at intake	Water level increase in the river at intake	Large diversion of sediment at the intake
	(2) Uncertainty	The extent of the increase in water level	The extent of the increase in water level	The extent of sediment intrusion
2. What is flexibility required for?	(3) Goal	To prevent high flood diversion and minimize damage	To manage flood peak diversion and maintain irrigation	To accommodate peak flood and maximize irrigation
		Avoiding costly interventions	Reducing cost of current intervention	Reducing the cost of future intervention

3) What are the dimensions of flexibility?

Capabilities are the system characteristics that enhance the system's flexibility to make necessary adjustments to cope with change and uncertainty. Capabilities are described by the sub features (DiFrancesco and Tullos, 2014): redundancy, achievability, coordination, connectivity, and slack. The characteristics that determine the system capabilities differ by system type (as summarized in Table 5-4).

Redundancy refers to the total number of options that can be achieved to meet future needs (Volberda, 1996, Gerwin, 1993, Slack, 1983). In the spate irrigation context, redundancy is provided by the number of intakes, bypass canals, flood depression zones and surface storage options; the number of organizations, or agencies for delegation of responsibilities such as maintenance and financing; and the number of alternative decisions on water sharing period and irrigation plans. For example, in Iran a number of reservoirs and individual terraces were built to accommodate different flood sizes (Kowsar, 2011a). In Myanmar a pump system along the river banks was implemented to pump excess water to supplement spate irrigation canals when needed (SIN, 2013a). In Pakistan and Yemen, improved traditional systems can use a number of options to cope with high peak floods such as bed stabilizers, flow dividers, and reinforcement of earthen structures (Steenbergen *et al.*, 2011).

Achievability (or agility) refers to the ability to implement a measure rapidly in a cost effective manner. The use of simple and less complex technologies is considered one of the advantages of traditional spate irrigation systems, supporting timely decisions and rapid actions such as to relocate intakes and structures or rebuild with local materials (Steenbergen *et al.*, 2011, Haile *et al.*, 2006). In the GAS (Sudan) water managers can make quick decisions about start of the irrigation period based on observations of the river (Fadul *et al.*, 2019). In acute water scarce situations they can decide to allocate water to those farmers who paid their water fees or have done field preparations, excluding those who are not ready (Fadul *et al.*, 2019). Other forms of achievability includes established mechanism to access to climatic, hydrologic, flow measurements and irrigation data; and institutional arrangements to maintain skilled labour for operation and maintenance of complex systems.

Coordination refers to the intra-basin coordination and system coordination between water managers and agencies related to the operation and management of water resources and flood management with due attention to regional coordination. For example, in Sudan, during the flood season, coordination between the River Training Unit in the Ministry of Water resources and the irrigation authorities in the improved traditional system of the GAS contributes to timely and appropriate operational decisions during high peaks and untimely floods and facilitates maintenance activities by borrowing heavy equipment (Fadul *et al.*, 2019). In traditional systems, cooperation between upstream and downstream farmers is essential for the jointly constructing and rebuilding of weir and diversion bunds (Haile *et al.*, 2011). Modernized systems hardly make use of farmers' cooperation for operation and maintenance which may have contributed to the failure of some modernized systems in the Raya valley in Ethiopia (Castelli *et al.*, 2018).

Connectivity refers to the ability of system components to attach to other components inside and outside the system (DiFrancesco and Tullos, 2014). In spate irrigation, conjunctive use of groundwater and surface water, infiltration basins and watering ponds are some forms of connectivity. Conjunctive use of canal water and groundwater enhance flexibility of irrigation service (Steenbergen *et al.*, 2011, SIN, 2013a). In Hadramut (Yemen) the use of sub-surface dam and low-level weir for groundwater recharge was one of the objectives of spate water management (van Steenbergern *et al.*, 2011).

Slack refers to the ability of a system to provide surplus capacity to cope with changes and uncertainty. Having excess capacity to divert, convey and store excess flood water creates opportunities to benefit from an increase irrigated area, increased urban water supply and groundwater recharge. (DiFrancesco and Tullos, 2014) argue that provision of slack in the form of bypasses and spillways provides a more cost-effective means to prevent flood damage than reinforcement and heightening interventions. Slack in spate irrigation system could be described by the degree of excess capacity that allows and give room for future adjustments to cope with changing conditions. A common provision of slack for coping with rising water levels is the option to delay interventions until uncertainties unfold over time (Anvarifar *et al.*, 2016, Fadul *et al.*, 2019). An example of this the choice for increasing the width of the embankment to allow for future heightening, or the provision of space around the embankment for future decisions on widening and heightening (Woodward *et al.*, 2011). In the traditional spate irrigation system in Baluchestan (Iran) slack is provided by increasing the storage capacity through a stone wall dam constructed across a narrow valley and connected with spillway. The wall is protected against excess high floods by the provision of drainage through weep holes (Kowsar, 2011a). In a modernized spate system in Iran, slack is provided by the construction of tail-end drains, spreading excess flood water over an extended area and using of groundwater recharge basins (Kowsar, 2011a). To handle high peaks excess flood water is sometimes diverted to forests and rangeland (Van Steenbergen *et al.*, 2010). For example, in Iran excess flood water diverted to orchards yielded 20,000 ton of figs (Kowsar, 2011a). During low flood events, slack is about finding additional water sources such as conjunctive use of groundwater, field water harvesting, and sharecropping.

The quantitative assessment of a system's capability requires metric descriptions of each sub-component of flexibility (i.e. redundancy, achievability, coordination, connectivity and slack), an example of which can be found in (DiFrancesco and Tullos, 2014). In this research we used a general description and assumptions to illustrate the application of the framework. Quantification is out of the scope of this research.

Table 5-4: Capabilities of different spate irrigation systems

Capabilities		Traditional system	Improved traditional system	Modernized system
1-Redundancy		High: a number of intakes, a number of fields (field and field system)	Very high: a number of intakes, a number of field to field system, a number of individual field system	Poor: one intake
2-Achievability	Simplicity	High: easily rebuilt, maintained, relocated; use of local material and knowledge; farmer-managed systems	Medium: easily built; farmer with support of local government management system	Low: Complex system; local government in partnership with farmers agency management system
	Cost	High maintenance cost including labour input	Average	Low maintenance cost, high initial cost
	Information system	Less dependent	Moderately dependent	Moderately to highly dependent
	Human power	Largely dependent on the number of farmers	Dependent on experienced farmers and engineers	Less dependent on farmers, highly dependent on qualified engineers and labours

3-Coordination	Cooperation between u/s and d/s farmers	Farmers-water managers-governmental agencies, upper catchment	Water managers-governmental agencies, upper catchment
4-Connectivity	Highly connected to ecosystem & groundwater recharge	Highly connected to ecosystem & groundwater recharge	Highly connected to ecosystem & groundwater recharge
5-Slack	Low: built to withstand only average and low floods	Average: spillways, scope to increase embankment and structure height	High: storage options and excess capacity

Mode of response describes the attitude of decision makers based on the effects of change (Golden and Powell, 1999); before (proactive), during, and after (reactive) occurrence of events. During and after events actions are event-driven aiming at reducing the negative consequences of impacts and losses (Evans, 1991), while actions and decisions before the event (proactive) anticipate external changes by taking measures to prevent the negative impacts and pursue possible opportunities (Triantis, 2000). (Fadul *et al.*, 2019) discuss the measures and decisions taken before, during and after events in response to flood risks by water managers, water user associations (WUA) and farmers in a spate irrigation system in Sudan. The mode of response of water managers and WUA were mostly before and during events (proactive), however farmers' response were mostly after the events (reactive) reflecting the risk averse behavior in subsistence farming systems (Table 5-5).

Temporal dimension indicates the period of time during which the change need to occur (De Toni and Tonchia, 1998). There are three categories for the temporal dimension of flexibility: strategic, tactical, and operational (De Neufville and Scholtes, 2011). Operational flexibility concerns timely and quick reactions to short-term, discrete changes that occur during or after an event. Examples of such events in spate irrigation are: a sudden rise in flood peak, overtopping of diversion intakes, breaching of river and canal embankment. Operational flexibility means that operations can be handled within system capacity without major setup (Stevenson and Spring, 2007, De Toni and Tonchia, 1998). For example, in spate irrigation, operational flexibility in response to a sudden rise in flood peak refers to quick changes in the operation of intakes and canal offtakes to avoid peak floods entering the canal system, or the diversion of peak floods away from the system to an infiltration basin. Tactical flexibility concerns short-term (within season) occasional changes that require some efforts and commitments without major changes in system setup (Anvarifar *et al.*, 2016, Rees, 1987). Examples of tactical changes in spate irrigation are the reduction of irrigation application time, reduction in irrigated area in case of low floods or drought conditions, changes in field layout or field canals. Strategic flexibility concerns medium or long-term changes such as change in irrigation requirements (Palmer and O'keefe, 2007), land entitlements, constructing new canals or irrigation structures, change of irrigation source, and abandonment decisions.

Table 5-5: dimensions of flexibility for different types of spate irrigation systems

Flexibility question	Flexibility characteristic features	Traditional system	Improved traditional system	Modernized system
3. What are the dimensions of flexibility?	(5) Temporal	Change is temporary, and short-term	Change is gradual, discrete and short-term	Change is permanent, and long term
	(6) Mode of response	A reactive response, after event	A reactive response during and after event	A proactive response

4) What needs to be changed or adapted?

Real options refer to a group of actions and managerial decisions for *adjustability* as a response to change and uncertainty (Myers, 1977, Triantis, 2003). It includes options to adapt as well as options to cope (DiFrancesco and Tullos, 2014). The concept of real options was first introduced by Myers (1977) which means the 'right, but not an obligation' to change/modify any system to adapt to changing environment (De Neufville and Scholtes, 2011, Anvarifar *et al.*, 2016). Real options are distinct from alternative choices by having the possibility to revise a decision at any time (de Neufville, 2002). Decision making in spate irrigation involves high uncertainty regarding flood volume, and duration before and during the irrigation season, which calls for an adaptive approach consisting of real options because future outcomes are not known. Real options allow for multiple decisions made over time as more information becomes available on events and impacts (Linquiti and Vonortas, 2012). Inflexible decisions (non-real options) use a deterministic approach for designing optimal strategies, based on a single fixed design value (such as diversion intake level, embankment height). Many designs of modernized spate irrigation use conventional design methods based on deterministic forecasts of the most likely scenarios or empirical methods, often followed by improvements in response to changes in river bed level and economic development. System *capabilities* should be able to support the actions and decision to cope to a changing situations. The adjustability of actions is the ability to expand, defer or shrink (Trigeorgis, 2005). Adjustability of decisions are options to delay or update. Adjustability to adapt refers to the ability to add, modify or remove any component of the system with ease. An important consideration in the flexibility framework is the engagement of farmers in the solution for the change. The decisions and actions with regards to adjustability must involve farmer's participation in the development of any alteration such as adding a new or improved structures and making managerial decisions to add, update or delay. Considering farmers' preferences allow the integration of best traditional practices in the technical design of new or improved options (Castelli *et al.*, 2018).

Enablers or flexibility mechanisms (Mikaelian *et al.*, 2011) refer to the facilitators or the enabling environment necessary for adjustment to take place.

5.3 FLEXIBILITY AND REAL OPTIONS IN TIMES OF HIGH, LOW AND UNTIMELY FLOODS

In the following section we describe the application of the developed framework to assess the flexibility of different types of spate irrigation systems to cope with different flood types such as high peak flood, low peak flood, and untimely peak flood.

5.3.1 High flood event

The results of flexibility assessment of traditional, improved-traditional and modernized-systems during a high peak flood event, are listed in Table 5-6.

Table 5-6: Flexibility and real options to cope with high floods in different types of spate irrigation systems

Flexibility question	Flexibility characteristic features	Traditional system	Improved traditional system	Modernized system
4. What needs to change or adapted?	(7) Real options (strategy)	Rebuild and/or change location of intakes and weirs	Expansion of existing structures (increasing height and/or width) (Figure 5), Decisions on opening or closing gates	Add new structures such as sediment control, trash control (Figure 6)
	(8) Enablers	Use of local materials, availability of farmers' knowledge and skills, and community action	Availability of improved materials, knowledgeable water managers, inclusion of farmers' experience	Building materials, outside support (government), engineering know-how, and inclusion of farmers' knowledge

In traditional spate systems, high peak floods leads to the collapse of upstream weirs and earthen bunds. Although traditional systems have several diversion intakes along the river to enhance the possibility of irrigating at least some of the area (Van Steenbergen, 1997), farmers are still uncertain about the failure of the downstream intakes. They are also uncertain regarding the extent of and recurrence of high floods. The flexibility of traditional systems stems from their simplicity, easy of adjustability in changing layout, moving the location to a more stable section or even making a new diversion weir using locally available material. Because of these flexibility features, real options to cope with high floods are related to timely decisions and actions of rebuilding or relocating intakes to catch some of the later floods. The flexibility to rebuilt or relocate damaged intakes, using locally available material, is to a large extent dependent on participation of all famers (enablers). For example in Myanmar, community participation resulted in the increase of irrigated area in dry zones (SIN, 2013a). However, the cost of rebuilding flexibility is high due to high labour and resource requirements.

Improved traditional systems are more robust in managing high floods than traditional systems, by using gated and reinforced structures to protect the command area from damage during large and uncontrollable floods. However, sudden rises in high water level and peak flows can result in failure of embankments of the river and canals, and in high pressure on the system due to lack of clear decision making in gate operation (Fadul *et al.*, 2019). Changing the size of structures and embankments provides slack and the availability of several intakes provides redundancy to secure diversion to at least some part of the area. Being able to adjust operational decisions on the diversion of potentially destructive high floods enhances system flexibility and creates real options to cope with uncertainty. For example, in eastern Sudan, an improved traditional diversion system in the GAS can irrigate up to 100,000 hectares of command area using gate control intakes, a spur system to stabilize channel bed, and river bank stone pitching. The operational plan in GAS is based on closure of intakes during passage of very high flood peaks. Although weak river embankments are occasionally exposed to failure, this also provides opportunities to some unplanned area to be irrigated. In Morocco, triangular dissipation structures are constructed using successive gabion spillways to allow safe distribution of water (Oudra, 2011). In Iran, bank stabilization using brush wood and river bed stabilization using gabion aprons, are common (Kowsar, 2011b). Gabion structures in the Punjab (Pakistan) provides flexible, and cost effective option for bed stabilization (Oosterbaan, 2010). In the Harosha spate system in Ethiopia, farmers block very high floods from entering the irrigation system by insertion of earth piles at the intakes and building fuse appendices at the end of diversion structure (Castelli *et al.*, 2018).

Being most robust, the modernized system can handle high peak floods, though this may encourage morphological changes due to erosion or build-up of sediment at upstream diversion bunds (Steenbergen *et al.*, 2011). The capacity of the field system is influenced by canal sedimentation and requires high initial investment cost (Steenbergen *et al.*, 2011). Additionally, water managers are uncertain of the extent of sediment accumulation in the system. The high sediment load entering the intake and canals during high flood events is a big concern for modernized systems, leading to degradation of irrigation infrastructure, poor water distribution and system failure (Castelli and Bresci, 2017, Libsekal *et al.*, 2015, Van Steenbergen *et al.*, 2010, Van Steenbergen, 1997, Haile *et al.*, 2011, Oosterbaan, 2010, Komakech *et al.*, 2011). Modernized systems tend to make adaptive, proactive decisions to anticipated changes through structural measures, without considering farmers' needs and established rules. Incorporation of indigenous practices and local knowledge contribute substantially to the rate of success and sustainability of adjustability decisions and actions, and hence flexibility. Failures in the modernization of traditional system in Eretria were caused by wrong estimates of design flood and scour depth (Van Steenbergen *et al.*, 2011), as well as lack of consideration of the existing rules and practices in the design and operation (Haile *et al.*, 2005a). Modernized system are accused to enhance unfair water distribution by favouring the upstream farmers with more water access (Haile *et al.*, 2011). However, when well managed, they provide minimum maintenance and routine work, allow flexible regulation, and adopt hydraulically effective structures (Geleta, 2014).

Figure 5.5: Intake heightening in improved traditional system in Sudan-Gash scheme. Photo by Eiman Fadul

Figure 5.6: Trash deflector in Yemen. © Ahmed Al-Siddig

5.3.2 Low flood event

The application of the flexibility framework for traditional, improved traditional and modernized system during a low floods are described in Table 5-7.

Table 5-7: Flexibility and real options to cope with low floods in different types of spate irrigation systems

Flexibility question	Flexibility characteristic features	Traditional system	Improved traditional system	Modernized system
4. What needs to change or adapted?	(7) Real options (strategy)	Decisions to modify cropping pattern or crop variety, reduce irrigated area, share risk Actions to augment water (groundwater)	Decision to update the irrigation plan: divert water to easily accessed fields, reduce irrigated area, field preparation (summer tillage)	Actions to add water from alternative sources: shallow groundwater wells, maximize diversion from the river, recharge wells
	(8) Enablers	Farmers' autonomy to choose crops & cropping pattern, availability of alternative water sources	Knowledgeable water managers and farmers willing to change plans	Availability of groundwater

Low floods and river water levels in traditional spate systems lead to water scarcity conditions and hence in reductions of the irrigated area, deficit irrigation and low crop production. Dealing with low flows involves decisions on water allocation, distribution and size of area to be irrigated. For example, farmers can decide to deal with low flows by diverting all available river water, prioritizing the upstream intake and area (Van Steenbergen, 1997). Farmers tend to increase their options by augmenting available water through water harvesting and conjunctive use of shallow wells; change crop choice; share risks through sharecropping and digging small ditches to distribute water flow (redundancy).

In improved traditional systems, during low flows water managers are uncertain how much area can be irrigated and hence tend to adjust the irrigation plan. To spread the risk of water scarcity among farmers, in the GAS (Sudan) water managers introduced a lottery system to allocate fields who receive water first. In the next year the order of the water allocation is reversed. Similar adjustability measures in improved and traditional systems are taken at field level. For example the use of lottery system to allocate actual irrigated area between farmers with priority given to un-chanced farmers from last season (Fadul et al., 2019). One of the adjustability actions in the spate system around Harosha river in Ethiopia is the construction of a bund to intercept low flood channel (Castelli et al., 2018).

In modernized systems, the concern of water managers during low floods is the reduced water diversion and uncertainty about the area that can be irrigated. During low floods, modernized systems can hardly divert water through intakes due to tendency of the main river channel to flow away from intakes (Hiben and Tesfa-Alem, 2014b). Adjustability options to enhance flexibility may include seeking other water supply sources to supplement irrigation in addition to famer's measures at field level.

5.3.3 Untimely flood event

The application of flexibility framework on traditional, improved traditional and modernized system during untimely flood events is described in Table 5-8 below.

Table 5-8: Flexibility and real options to cope with untimely floods in different types of spate irrigation systems

Flexibility question	Flexibility characteristic features	Traditional system	Improved traditional system	Modernized system
4. What needs to be changed or adapted?	(7) Real options (strategy)	The decision to modify crop, share risk, action to rebuild the bund	The decision to delay the irrigation schedule, action for supplementary irrigation	The decision to modify maintenance plan to start early, action to supplementary irrigation
	(8) Enablers	Farmers' knowledge about crops, group decision making	Availability of participatory decision making system including farmers	Availability of resources for maintenance activities, participatory decision making including farmers

In traditional systems, the change in the starting date of the season results in great uncertainty about the extent and length of the flood duration period due to the implications on the optimum cropping dates. Flexibility is provided through adjustable managerial decisions on handling the risk at field level and actions to rebuild bunds that depend on use of local materials such as trees, shrubs and soil to protect fields and bunds.

In improved traditional system, with similar concerns of change and uncertainty, water managers adopt real actions for adjustability such as an emergency maintenance to the most critical structures and embankments and borrowing maintenance equipment from other collaborating agencies (cooperation) such as in GAS (Fadul et al., 2018, Fadul et al., 2019), and conjunctive use of groundwater as a supplementary source (redundancy). This option provided farmers in Wadi Zabid in Yemen with a substantial increase in irrigated area (Van Steenbergen et al., 2010). Options for groundwater recharge provide linkage between spate irrigation and natural resources management (connectivity) such as in Wadi Hadramout in Yemen as described by Van Steenbergen et al. (2010).

In well managed modernized system decisions can be more sustainable with appreciated efforts for maintenance and preparation well before the expected season (Slack) and conjunctive use of ground water (Redundancy). Provision of flexibility at farmer level is similar to the other spate systems and related to managerial decisions such as change crop.

5.4 Conclusion

Spate irrigation systems face a high level of uncertainty using an unpredictable water source from flash floods in ephemeral rivers. Flexibility is a key ingredient of coping strategies. In this paper we developed a framework to assess flexibility in spate irrigation systems. The framework is based on earlier work by (Anvarifar *et al.*, 2016) and DifrancescoTullos (2015), applied to three types of spate irrigation systems: traditional, improved traditional and modernized systems.

Traditional systems were considered as the most flexible, followed by improved ones. Modernized systems with fixed concrete structure were considered the least flexible. The results of flexibility assessment showed that traditional systems can cope with changes and avoid damage from destructive high flood peaks or untimely floods by selecting real options to rebuild or relocate intakes and diversion weirs. During low floods, real options available at field level include changing crops or reducing irrigated area. Similarly, improved traditional adjust to high flood peaks with real options such as timely closing of gates at the intakes and diversion weirs. During low flood and untimely floods, improved systems adjust through reduction in the irrigated area and adjustments in irrigation scheduling, respectively.

Being considered least flexible, modernized systems require high capital cost to implement real options such as constructing stilling basin to deal with sediment laden flows during peak floods. However, in several countries the management of modernized systems lack farmer's involvement in decision making and proposed actions, which resulted in failures (Castelli *et al.*, 2018, Van Steenbergen *et al.*, 2010). The requirements for the resources for maintenance are often underestimated and the systems failed because of poor maintenance (Castelli *et al.*, 2018). Low floods can be dealt with by reducing the irrigated area and adjustments in the water sharing rules. These options are only successful with farmers' involvement in the decision process. Ignoring farmers' input leads to the limited number of sustainable options for modernized systems to cope with low flood events. Therefore modernized systems are able to provide more durable and sustainable irrigation system if managerial decisions are adjusted to include participatory approach which was also confirmed by a pilot study conducted by (Castelli *et al.*, 2018).

It can be observed from the global experience that the flexibility of improved systems is relatively the best compared to traditional and modernized systems, in terms of the damage and benefit on the system. HibenTesfa-Alem (2014b) compared traditional and modern spate systems in Tigray (Ethiopia). The authors attributed the failure of the modernized system to the underestimation of the design water level and sediment trend.

It is challenging to solidly attribute the high failure rate of modernized systems to their lack of flexibility as described by Steenbergen *et al.* (2011), HibenTesfa-alem (2014a), CastelliBresci (2017), Libsekal *et al.* (2015), Van Steenbergen *et al.* (2010), Van Steenbergen (1997), Haile *et al.* (2011), Oosterbaan (2010), Komakech *et al.* (2011). The limited knowledge on design and management of complex structures to handle unpredictable high floods with large sediment load and the poor use of traditional farmers' knowledge may have contributed to failure. Yet, global experiences showed that although modernized system were effective in some cases, improved traditional systems were generally more effective. In fact, modernized system can become more flexible by adapting operational plans to avoid peak flood diversion, improving rules for equal water distribution and, most importantly, involving farmers in decision making regarding operation and maintenance activities.

The comparison of flexibility of different systems reveals that all spate irrigation systems can maintain flexibility through exploring and implementing a number of real options which could serve during occurrence of risky events. Failure of some of the modernized system around the world can be avoided by exploring adjustable managerial decisions that includes farmers and more participatory approach in the design, construction, maintenance and operation of the system. Farmers and water managers can manage flood variability more successfully when a range of options or alternative paths are available. The framework provide a range of measures that integrate flood risk management (high, low and untimely) and irrigation development.

Conjunctive use of groundwater and spate irrigation to supplement irrigation during low and untimely floods enhances flexibility of all types of spate irrigated systems, while the impact on groundwater level can be reduced by groundwater recharge wells and basin as practiced in many countries.

The conceptual framework serves as a professional guide for policy maker, water managers, WUAS, and farmers to make preliminary evaluation of their system, explore goals, capabilities and options available. For policy makers, they will be able to determine the level of investments they need to take in order to achieve sustainability and prosperity of the livelihood of spate irrigation communities. For water managers, they will be able to assess different options available, constraints and limitations for development and enhancement. For example, if a certain system lacks slack, then provision of access to fuse blogs, free board or natural drainage system could be explored in a proactive manner. Similarly, farmers will be able to explore different options to go further for implementation.

It is outside of the scope of this paper to quantify flexibility of the three system types. This could be done by quantifying the sub-features describing the capabilities as shown by DifrancescoTullos (2015) to compare systems types and prioritize potential options (Gupta and Goyal, 1989).

6

CONCLUSIONS AND RECOMMENDATIONS

This chapter presents the conclusions and recommendations drawn from this research. These are provided after presenting a brief summary of the main results and the limitations associated with this research.

6.1 GENERAL

Through this PhD thesis, assessment of risk and coping strategies for managing highly variable and uncertain water supply in spate irrigation systems, were presented through development of different methodologies and frameworks. The aim of this research was to assess the risks and coping strategies to cope with uncertain water supply in spate irrigation to contribute towards achieving sustainable livelihood farming communities, taking the Gash agricultural scheme (GAS) in Sudan as a case study. With this aim, a number of sub-objectives were defined: 1- to study the main elements of uncertain water supply risks that have significant impact on irrigation performance, 2- to evaluate the effectiveness of coping strategies and practices that have been developed over years to cope with uncertain water supply, 3- To identify alternative locally feasible measures that would address the different level of hydrological events and cope with variability of water supply and enhance irrigation performance, and 4- to establish a conceptual framework for adoption of real option that enhance system flexibility to cope with variability and uncertainty of water supply.

To achieve these objectives, different methodological frameworks were developed for risks and coping strategies assessment:

- The Source-Pathway-Receptor-Consequence Model was adapted and developed for analysing and assessing risks of using unpredictable flash floods as a source of irrigation in spate-irrigated agriculture. Risks were investigated using farmers', WUAs' and water manager's perceptions on risk categories, pathways and consequences for different stakeholders at spatial level including upstream, midstream and downstream locations. Pathways included stakeholders' perceptions on risks, flood variability in terms of volume, duration and timing, infrastructure, operation and maintenance, and institutional arrangements.

- The Driving force-Pressure-State-Impact-Response (DPSIR) framework was used to establish the cause-effect relationships between the water supply variability and stakeholders' coping strategies. This helped in problem structuring, evaluating the effectiveness of different flood management strategies and for the development of water strategies contributing to sustainable resources management. Additionally, the mDSS4 (The MULINO Decision Support System) tool facilitated the involvement of stakeholders in the process of Integrated Water Resources Management and natural resources management. For example identification of the existing response measures during high, low and untimely floods within a cause-effect environment, evaluation of the effectiveness of the coping strategies, selection of the evaluation criteria and indicators, and choosing the most effective measures for low, high and untimely flood strategies that perform better with respect to the selected criteria.

100

- Surface irrigation modelling using WinSRFR model was used to evaluate performance of current field design and alternative field designs under different application time (T_{co}) during large, medium and small flood seasons. Three strategies were investigated: 1) time management strategy for current field design; 2) time management strategy for an improved (alternative) field design, and 3) flow management strategy for an improved (alternative) field design. The current field design was a large sized-border field (8,400 m×500 m), while the improved field design was the alternative design based on vertical or horizontal division of current field layout i.e. (8,400 m×250 m) or (4,200 m×500 m). In the first strategy, irrigation water was conveyed to the whole current field layout at different application times. In the second strategy, the total application time was divided equally between two sub-divided fields. While in the third strategy, total inflow was equally divided between two sub-divided fields. Irrigation performance of different combinations of flood size, field layout and application times were examined using application efficiency, distribution uniformity, and adequacy criteria to obtain the best performing scenario.

- Effective coping strategies in spate irrigation were the real options to cope with uncertainty and variability of hydrological event described by flexible approaches in the actions and decision. Therefore a novel approach for establishment of real options in spate irrigation systems was developed. The framework consisted of four principle questions, and eight main flexibility features and five sub-features found to represent flexibility in spate irrigation system based on relevant literature. The conceptual framework demonstrated its beneficial use for the evaluation of spate irrigation system through its application on traditional, improved traditional, and modern spate systems to cope with high peak floods, low peak floods and untimely flood events. The conceptual framework could also serve as a professional guide for policy maker, water managers, WUAS, and even farmers to make preliminary evaluation of their system, explore goals, capabilities and options available. For policy makers, they would be able to determine the level of investments they need to take in order to achieve sustainability and prosperity of the livelihood of spate irrigation communities. For water managers, they would be able to assess different options available, constraints and limitations for development and enhancement. Similarly, farmers and WUAs would be able to explore different options to go further for implementation

6.2 CONCLUSIONS

A number of conclusions could be drawn and formulated in relation to the specific objectives described above:

6.2.1 Main sources of risk

Observations of flood events in the historical records of hydro-climatic data were categorized based on stakeholder's perceptions on threshold values. The main sources of risks were: low flood, high flood, short flood, extended flood, early flood and late flood. Findings showed that farmers, WUAs and system managers perceived the risks from floods differently. Further the source of risk, pathways and consequences were dependent on the location of farmer, WUAs and water manager in the system with more impacts on downstream locations. The survey revealed that the experience of farmers with water supply failures for their individual fields had strongly influenced their perceptions about the risks posed by different types of flood events. The farmers were primarily concerned by low floods had a tendency to underestimate high floods, while the WUAs were more disturbed by untimely floods. The system managers were most troubled by high and potentially destructive floods. The poor state of the infrastructure, lack of proper maintenance and suboptimal operation aggravated the consequences of water supply risks. Consequently, the impacts were low crop yield, highly variable crop production and highly variable irrigated area. Therefore upgrading of physical infrastructures, and improving/updating the policy and institutional support for the GAS could be one of the pillars to increase the capacity to manage risks due to uncertain water supplies

6.2.2 Coping strategies and adoption

The unpredictability of floods lead to uncertain and unequitable irrigation water supply due to low, high, and untimely flows into the irrigation system. The field survey differentiated between flood risk categories and the different measures used to cope with them, in particular, low flood strategy, high flood strategy flood, and untimely flood strategy. The local measures were assessed and ranked based on identified environmental, management, social and economic criteria:

- Low flood strategy: The farmers and WUAs developed a larger number of medium and highly effective measures for low flood strategy than water managers. The most effective measures to cope with low floods were proactive actions such as land and soil preparations by farmers, and mesquite clearance by WUAs, and mapping of flooded area every 10 days by water managers. The analysis revealed striking differences in effectiveness scores according to the location and type of stakeholder. All effective measures could be implemented by farmers and WUAs, and WUAs without external support technical difficulty. However, most effective measures were not well adopted by all the stakeholders because of limited access to financial resources and supportive institutes and policies.

- High flood strategy: The highly ranked measures were reactive managerial decisions for farmers such as delaying the start of cropping date and field access, and change to water-intensive crops. Meanwhile, the highly ranked measures were proactive decisions and actions for WUAs and water managers such as desilting of irrigation canals, raising field embankments and routine maintenance. Yet the majority of farmers' and WUAs' measures were not among the highly effective ones since the adoption level depends on the capacity, available resources and support systems. Field preparation, mesquite clearance and maintenance activities for irrigation canals and embankments required financial and institutional support which were not available to the majority of farmers and WUAs. Unlike WUAs and farmers, the water managers were better equipped for dealing with the high floods. Almost all the measures adopted by water managers were ranked as highly or medium effective in dealing with high floods since irrigation infrastructures were more impacted by high water level. Nevertheless, adoption of measures was location-dependent.

- Untimely flood strategy: The number of measures developed for untimely floods were few with low adoption of effective measures by all stakeholders due to limited resources and poor institutions. If untimely floods occurred during the maintenance and preparation activities, this would lead to breaching of the poorly maintained canals and structures. Effective measures were sediment and flow energy control measures using local vegetation for farmers, lottery system for land distribution after irrigation for WUAs, and involving private sector in maintenance activities for water managers.

6.2.3 Performance of alternative measures

- Under time management strategy, the performance was assessed for the current field design with large sized-border field (8,400 m×500 m), under different application time during high, average and low floods. This strategy showed poor optimal performance of DU, AD, AE indicators under large, medium, and small flood season. In general, the optimum performance showed over irrigation and poor DU during large floods, and under irrigation and poor DU during medium and low floods. The results showed that reducing 30% of the field area at the tail end could save 60% of the application time under average inflow rates and could reduce 50% of water losses to deep percolation. Implication of this could improve the current practice through increase of irrigation efficiency at head-field locations since further irrigation of tail-end fields using the current upstream downstream approach resulted in increased deep percolation losses. Additionally, during low flow conditions, reducing the field area could significantly increase the chance of irrigation of the upper-head field located further downstream the irrigation offtake.

- Under time management and improved field design strategy, performance was assessed using vertical division (8400×250m) and horizontal division (4200×500 m), different flood sizes and divided application time (0.5 T_{co}). Performance resulted in higher indicator values compared to other strategies. Further, the results revealed similar performance of vertical and horizontal improvements on field design during large, medium, and small flood seasons. This strategy achieved high performance indicators compared to current field design under different application times and flood sizes. The proposed strategy could save 40% of the current application time during large flood seasons, and 20% of the time during medium flood seasons.

- Under flow management and improved field design strategy, performance was assessed using divided inflows (0.5Q) diverted to each sub-field, and application time. The results revealed better performance of horizontal sub-fields during high, and average flood seasons, and poor performance during performance of both vertical and horizontal sub-fields during low flood seasons. Findings from the analysis of third strategy were similar to first strategy since performance of total flows into total area with similar application time was the same. Poor DU and over-irrigation results were observed at the optimum condition. Higher AE is found to be associated with poor irrigation performance of AD and DU, particularly at small application time.

- This research proved that field size and application time were the main elements which had substantial influence on irrigation performance. Farmers experience on field division without a scientific evidence for validity proved to work successfully, therefore farmers coping strategies should be considered when conducting applied research.

6.2.4 Real options to cope with uncertainty and variability

Traditional, improved-traditional and modernized spate irrigation systems could maintain flexibility if it was well planned and included in advance through exploring and implementing a number of real options/coping strategies that could serve under risky flood events. The condition of making adjustable managerial decisions and actions that include farmers using a participatory approach was vital to maintain flexibility of spate irrigation systems. Farmer involvement in the design, construction, maintenance and operation of the system, enhance system flexibility to cope with variability through exploring/provision of a range of options or alternative paths. The framework provided a range of measures that integrate flood risk management (high, low and untimely) and irrigation development such as rangeland developments, groundwater recharging basins, and conjunctive use of groundwater and spate irrigation to supplement irrigation during low and untimely floods. This could increase irrigated area and enhance flexibility of all types of spate irrigated systems. The impact on groundwater level could be reduced by groundwater recharge wells and basins as practiced in many countries.

6.3 MAIN CONTRIBUTION

This research has high scientific and development significance. Scientifically it contributes to more understanding of the methodologies required for assessment of risks and coping strategies to the uncertain water supply in spate irrigation in general, and in Gash Agricultural Scheme (GAS) in Sudan in particular. The research focused on flood risks such as high, low, and untimely floods and its impact on irrigation performance. Additionally, flood risk assessment approaches have been focusing only on urban system targeting protection of cities, towns and residential areas with high economic value. Risk assessment in low cost rural community system, such as spate irrigation system in arid and semi-arid zones, has been neglected in the literature.

Another scientific contribution is that the spate irrigation has been less recognized in the literature of irrigation technologies compared to other technologies. Additionally, few authors have discussed local cases focusing on system descriptions and recommendations for future development without detailed investigation on development of methodologies and scientific approaches for spate irrigation development.

On the societal aspects, the research contributed to development of improved locally feasible field design and operational rules that could improve irrigation performance and equity between farmers. A more effective field water distribution is expected to enhance food security and the livelihood of the farmers in spate irrigation system.

Another societal contribution was the analysis of risks at different spatial scales and for different stakeholders. This could help to formulate mitigation strategies to address the risks faced at different levels of the studied system.

The development of conceptual framework for flexible real options in spate irrigation system was an important contribution. It could serve as a professional guide for policy maker, water managers, WUAS, and even farmers to adopt effective actions and decisions to cope with variability and uncertainty.

6.4 RECOMMENDATION FOR FURTHER RESEARCH

Flood risk assessment helps farmers and policy makers in spate system to better cope with climate variability and to develop relevant adaptation strategies that enhance irrigation performance and hence crop productivity. Although this research shows the significance of risk and coping strategies assessment in spate irrigation systems, the following recommendations indicate some of the aspects for improving risk and coping strategies assessment in spate irrigation systems:

- This research conducted risk assessment without risk quantification. Since floods in spate irrigation are associated with both risks and opportunities. High flood damage in spate irrigation could be evaluated through determination of the damage cost as a function of a combined impact of an alarming water level upstream intakes, high water velocity in the river, local rainfall, extent and duration of high water level. Similarly, opportunities quantification should also be considered through quantification of gains that occur during an uncontrolled flood event such as the chance of increased irrigation and production of cash crops with high water requirement. Risk quantification should also include low flood events whereby damage cost determined by loss in crop production as a function of lowest acceptable water level upstream intakes, duration of low flood and input from local rainfall. Opportunities during low floods could be quantified through estimating the benefits of reducing the bets and diseases and mesquite control, and the more challenging to quantify is the social benefits of risk sharing during low floods. Therefore, further research to explore risk quantification could be an added value to the research on risk management in spate irrigation.

- Risk and coping strategies assessment highlighted the importance of policy and institutional support in increasing the capacity of farmers, water user associations and water managers to manage risks due to uncertain water supply. Further research on institutional arrangements and policy decisions needed to reduce or control risk impacts on farmers and the system should be explored.

- The use of multi-criteria decision making in the evaluation and scoring of the effective strategies need to be enhance with quantitative measures where possible to reduce the impact of subjectivity of decisions. Therefore further research on quantitative assessment of effective strategies is deemed essential.

- The use of simple low cost technologies and flexible decisions and actions by farmers and WUAs were mostly effective and highly adopted. Research on simple technologies or innovations for field design, sediment management, and water distribution at field level from other spate irrigation systems should be tested and explored in GAS.

- Selection of the infiltration parameters for the event analysis in WinSRFR model was obtained using trial and error of several empirical infiltration functions. A detailed soil survey analysis could provide the study site with more accurate representations of infiltration parameters. Therefore further soil analysis research to determine the site-specific infiltration parameters would strengthen the calibration and simulation results.

- The WinSRFR hydrodynamic model used in this study has some degree of uncertainty in the model structure and soil parameters assumptions which results in the model uncertainty. Therefore recommendation for field layout adjustments needs to be tested at pilot field for few seasons before upscaling. There is scope for further applied research to address those recommendations.

- The research presented a methodology to evaluate the flexibility of spate irrigation system using the capability features. A further research is needed to investigate on quantifying the capabilities features which can assist in exploring the added value of flexibility in spate irrigation systems.

APPENDICES

APPENDIX 3-I 1

Stakeholder	Low flood strategy	Ranking	Code	Score
Farmers	Land preparation before flood	1	SF-Low3	0.81
	Sharecropping	2	SF-Low8	0.74
	Use of shrubs and weeds	3	SF-Low14	0.70
	Summer tillage	4	SF-Low11	0.65
	Cultivate vegetables	5	SF-Low10	0.64
	Pre-tillage before flood season	6	SF-Low2	0.63
	Increase seeding rate for fodder production	7	SF-Low5	0.60
	Digging small ditches to distribute water flow	8	SF-Low13	0.59
	Double tillage	9	SF-Low4	0.56
	Wetting seeds to reduce the 1st growth developing stage	10	SF-Low6	0.55
	Reduction of cultivated area	11	SF-Low1	0.46
	Cultivate only on part of the field	12	SF-Low15	0.46
	Change sorghum variety	13	SF-Low9	0.46
	Social system of sharing benefits	14	SF-Low16	0.45
	Make small earth bunds	15	SF-Low7	0.44
	Exit cropping season (Do not cultivate)	16	SF-Low12	0.10
WUAs	Mesquite clearance	1	SWUA-Low7	0.72
	Temporarily land leasing to private sector	2	SWUA-Low13	0.72
	Laying shrubs & weeds at field head to dissipate flow energy	3	SWUA-Low10	0.71
	Change of water source to groundwater at head fields	4	SWUA-Low11	0.71
	WUAs at field level is divided in sub-groups of farmers for improved O&M	5	SWUA-Low14	0.70
	Flexible infield spurs for field water distribution	6	SWUA-Low3	0.68
	Participate in flood water spreading	7	SWUA-Low8	0.68
	Lottery system for field allocation to farmers	8	SWUA-Low1	0.66
	Fixed system for field allocation to farmers	9	SWUA-Low2	0.65
	Re-alignment of field canal	10	SWUA-Low5	0.64
	Longitudinal field division of irrigation fields	11	SWUA-Low12	0.61
	Manage irrigation period between adjacent WUAs	12	SWUA-Low4	0.45
	Sharing field canal between adjacent fields on different irrigation time	13	SWUA-Low9	0.41

	Allocation of one farm per farmer in every flooding season	14	SWUA-Low6	0.39
Water managers	Mapping of flooded areas every 10 days	1	SM-Low3	0.68
	Diversion of first floods to recharge groundwater & watering ponds	2	SM-Low6	0.59
	Allocation of fields with high and low chances of good irrigation for each WUA	3	SM-Low2	0.51
	Water allocation period with flexibility	4	SM-Low4	0.50
	Share maintenance burden with WUA to maintain secondary & field systems	5	SM-Low5	0.47
	Division of flood period in two irrigation schedules	6	SM-Low1	0.39

APPENDIX 3-I 2

Stakeholder	High flood strategy	Ranking	Code	Score
Farmers	Use of lebsha to dissipate flow energy D/S field intakes	1	SF-High2	0.81
	Use of sand bags for small breaches	2	SF-High1	0.74
	Cultivate water melon in winter	3	SF-High8	0.70
	Delaying the start time of cropping activities	4	SF-High4	0.65
	Cultivation of a second crop after harvest	5	SF-High10	0.64
	Close of water paths and gullies	6	SF-High6	0.63
	Double tillage to reduce weeding	7	SF-High9	0.60
	Fill the breach with shrubs and weeds (Lebsha)	8	SF-High3	0.59
	Change crop variety	9	SF-High5	0.56
	Breaching banks of nearby fields	10	SF-High7	0.55
WUAs	Field preparation (field canal desilting, heightening embankments, etc.)	1	SWUA-High1	0.91
	Report major breaching	2	SWUA-High4	0.80
	Laying shrubs and weeds at field head to dissipate flow energy	3	SWUA-High2	0.76
	Breaching embankments of adjacent fields	4	SWUA-High3	0.09
Water managers	Routine maintenance before flood season	1	SM-High2	0.84
	Raising offtakes of main canal and secondary canal	2	SM-High11	0.82
	Established water level gauges at intakes for effective operation	3	SM-High6	0.75
	River training works and strengthening of River embankments	4	SM-High1	0.71
	Cooperation with River monitoring units for early warning	5	SM-High5	0.69
	Use of labour to prevent accumulation of debris U/S offtakes	6	SM-High12	0.69
	Mobilizing financial resources and incentive system	7	SM-High113	0.69

Maintaining critical sections before flood	8	SM-High3	0.68
Start irrigation with first upstream and last downstream fields	9	SM-High7	0.66
Maintaining a reasonable distance between field offtakes	10	SM-High4	0.64
Flow releases to planned fields if canal stability is not threatened	11	SM-High8	0.62
Diversion of water into unplanned fields to release flow energy	12	SM-High9	0.41
Delay of maintenance work to the end of season	13	SM-High10	0.10

APPENDIX 3-I 3

Stakeholder	Untimely flood strategy	Ranking	Code	Score
Farmers	Use of sand bags for small breaches & seek assistance for major breaching	1	SF-untimely1	0.91
	Fill the breach with shrubs and weeds(Lebsha)	2	SF-untimely2	0.72
	Cultivate in winter	3	SF-untimely6	0.59
	Social system of sharing benefits by sharing irrigated fields or harvest	4	SF-untimely5	0.54
	Change crop	5	SF-untimely3	0.52
	Exit cropping season (Do not cultivate)	6	SF-untimely4	0.18
WUAs	Lottery system for field allocation to farmers	1	SWUA-untimely2	0.80
	Manage irrigation period between adjacent WUAs	2	SWUA-untimely1	0.45
	Use sand bags and seek assistance	3	SWUA-untimely3	0.05
Water managers	Involve private sector for maintenance activities	1	SM-untimely8	0.83
	Allowing flexible starting and end dates of irrigation	2	SM-untimely9	0.63
	Use of timber stop logs to control water level	3	SM-untimely4	0.58
	Priority of maintenance to WUAs who paid water fees	4	SM-untimely6	0.57
	Use of experienced gate operator to adjust openings	5	SM-untimely5	0.56
	Borrow maintenance equipment where possible	6	SM-untimely3	0.53
	Borrow from state government to deal with delay of budget	7	SM-untimely7	0.52
	Maintaining critical sections before flood,	8	SM-untimely1	0.52
	Emergency action on silt removal	9	SM-untimely2	0.36

REFERENCES

ABBASI, F., SHOOSHTARI, M. M. & FEYEN, J. 2003. Evaluation of various surface irrigation numerical simulation models. *Journal of Irrigation and Drainage Engineering,* 129, 208-213.

ABDELGALIL, E. & BUSHARA, A. I. 2018. Participation of Water Users Associations in Gash spate system management, Sudan. *Water Science,* 32, 171-177.

ADAMALA, S., RAGHUWANSHI, N. & MISHRA, A. 2014. Development of Surface Irrigation Systems Design and Evaluation Software (SIDES). *Computers and Electronics in Agriculture,* 100, 100-109.

ADELEKAN, I. O. & ASIYANBI, A. P. 2016. Flood risk perception in flood-affected communities in Lagos, Nigeria. *Natural Hazards,* 80, 445-469.

AGGARWAL, P. K., BAETHEGAN, W. E., COOPER, P., GOMMES, R., LEE, B., MEINKE, H., RATHORE, L. S. & SIVAKUMAR, M. V. K. 2010. Managing Climatic Risks to Combat Land Degradation and Enhance Food security: Key Information Needs. *Procedia Environmental Sciences,* 1, 305-312.

AIMAR, A. 2017. Managing Water Crisis in the North African Region: With Particular Reference To Jijel Region. *In:* BEHNASSI, M. & MCGLADE, K. (eds.) *Environmental Change and Human Security in Africa and the Middle East.* Cham: Springer International Publishing.

ALI, M. H. 2011. Irrigation System Designing. *Practices of Irrigation & On-farm Water Management: Volume 2.* Springer.

AMARNATH, G., SIMONS, G., ALAHACOON, N., SMAKHTIN, V., SHARMA, B., GISMALLA, Y., MOHAMMED, Y. & ANDRIESSEN, M. 2018. Using smart ICT to provide weather and water information to smallholders in Africa: The case of the Gash River Basin, Sudan. *Climate Risk Management,* 22, 52-66.

ANVARIFAR, F., ZEVENBERGEN, C., THISSEN, W. & ISLAM, T. 2016. Understanding flexibility for multifunctional flood defences: a conceptual framework. *Journal of Water and Climate Change,* 7, 467-484.

ANWAR, A. A., AHMAD, W., BHATTI, M. T. & HAQ, Z. U. 2016. The potential of precision surface irrigation in the Indus Basin Irrigation System. *Irrigation science,* 34, 379-396.

ASIF, M. & ISLAM-UL-HAQUE, C. 2014. Hill torrents potentials and spate irrigation management to support agricultural strategies in Pakistan. *American Journal of Agriculture and Forestry,* 2, 289-295.

AVELINO, J. 2012. *Optimization of farm water management and agronomic practices under spate irrigation in Gash Agricultural Scheme - Sudan* Master of Science, UNESCO-IHE Institute for Water Education.

AZUMAH, S. B., DONKOH, S. A. & AWUNI, J. A. 2018. The perceived effectiveness of agricultural technology transfer methods: Evidence from rice farmers in Northern Ghana. *Cogent Food & Agriculture,* 4, 1-11.

BASHIER, E. E., ADEEB, A. M. & AHMED, H. M. 2014. Assessment of water users associations in spate irrigation systems: case study of Gash Delta Agricultural Corporation, Sudan. *International Journal of Sudan Research,* 4.

BAUTISTA, E., CLEMMENS, A. J., STRELKOFF, T. S. & NIBLACK, M. 2009a. Analysis of surface irrigation systems with WinSRFR—Example application. *Agricultural water management,* 96, 1162-1169.

BAUTISTA, E., CLEMMENS, A. J., STRELKOFF, T. S. & SCHLEGEL, J. 2009b. Modern analysis of surface irrigation systems with WinSRFR. *Agricultural Water Management,* 96, 1146-1154.

BAUTISTA, E., S STRELKOFF, T., CLEMMENS, A. & L SCHLEGEL, J. WinSRFR: Current Advances in Software for Surface Irrigation Simulation and Analysis. 5th National Decennial Irrigation Conference Proceedings, 5-8 December 2010 2010 American Society of Agricultural and Biological Engineers and the Irrigation Association Phoenix Convention Center, Phoenix, Arizona USA. ASABE, 12pp.

BAUTISTA, E., SCHLEGEL, J., STRELKOFF, T. S., CLEMMENS, A. J. & STRAND, R. J. An integrated software package for simulation, design, and evaluation of surface irrigation systems. In World Water & Environmental Resources Congress, May 21-25, 2006 2006 Omaha, NE. , 10pp.

BAUTISTA, E., SCHLEGEL, J. L. & STRELKOFF, T. S. 2012. WinSRFR 4.1-User Manual. Cardon Lane, Maricopa, AZ, USA: USDA-ARS Arid Land Agricultural Research Center.

BILLIB, M., BARDOWICKS, K. & ARUMÍ, J. L. 2009. Integrated water resources management for sustainable irrigation at the basin scale. *Chilean Journal of Agricultural Research,* 69, 69-80.

BINSWANGER, H. P. & SILLERS, D. A. 1983. Risk aversion and credit constraints in farmers' decision-making: A reinterpretation. *The Journal of Development Studies,* 20, 5-21.

BO, C., ZHU, O. & SHAOHUI, Z. 2012. Evaluation of hydraulic process and performance of border irrigation with different regular bottom configurations. *Journal of Resources and Ecology,* 3, 151-160.

BORGA, M., ANAGNOSTOU, E. N., BLÖSCHL, G. & CREUTIN, J. D. 2011. Flash flood forecasting, warning and risk management: the HYDRATE project. *Environmental Science & Policy,* 14, 834-844.

BORGA, M., GAUME, E., CREUTIN, J. D. & MARCHI, L. 2008. Surveying flash floods: gauging the ungauged extremes. *Hydrological processes,* 22, 3883.

BOTZEN, W., AERTS, J. & VAN DEN BERGH, J. 2009. Dependence of flood risk perceptions on socioeconomic and objective risk factors. *Water resources research,* 45.

BROOKS, N. 2004. Drought in the African Sahel: long term perspectives and future prospects. *Tyndall Centre for Climate Change Research, Norwich, Working Paper,* 61, 31.

BROOMELL, S. B., BUDESCU, D. V. & POR, H.-H. 2015. Personal experience with climate change predicts intentions to act. *Global Environmental Change. ,* 32, 67-73.

BURT, C. M., CLEMMENS, A. J., STRELKOFF, T. S., SOLOMON, K. H., BLIESNER, R. D., HARDY, L. A., HOWELL, T. A. & EISENHAUER, D. E. 1997. Irrigation performance measures: efficiency and uniformity. *Journal of irrigation and drainage engineering,* 123, 423-442.

CAMACHO, R., 1987. Traditional spate irrigation and wadi development schemes. In: Spate irrigation: proceedings of the Subregional Expert Consultation on Wadi

Development for Agriculture in the Natural Yemen 6–10 December, 1987 Aden, PDR Yemen. UNDP/FAO, pp. 60–72.

CASTELLI, G. & BRESCI, E. 2017. Participatory rural appraisal for diagnostic analysis of spate irrigation systems in Raya Valley, Ethiopia. *Journal of Agriculture and Rural Development in the Tropics and Subtropics (JARTS),* 118, 129-139.

CASTELLI, G., BRESCI, E., CASTELLI, F., HAGOS, E. Y. & MEHARI, A. 2018. A participatory design approach for modernization of spate irrigation systems. *Agricultural Water Management,* 210, 286-295.

COOPER, P. J. M., DIMES, J., RAO, K. P. C., SHAPIRO, B., SHIFERAW, B. & TWOMLOW, S. 2008. Coping better with current climatic variability in the rain-fed farming systems of sub-Saharan Africa: An essential first step in adapting to future climate change? *Agriculture, Ecosystems & Environment,* 126, 24-35.

CORRAL-QUINTANA, S., LEGNA-DE LA NUEZ, D., VERNA, C. L., HERNÁNDEZ, J. H. & DE LARA, D. R.-M. 2016. How to improve strategic decision-making in complex systems when only qualitative information is available. *Land Use Policy,* 50, 83-101.

CREUTIN, J. D. & BORGA, M. 2003. Radar hydrology modifies the monitoring of flash-flood hazard. *Hydrological processes,* 17, 1453-1456.

CREUTIN, J. D., BORGA, M., GRUNTFEST, E., LUTOFF, C., ZOCCATELLI, D. & RUIN, I. 2013. A space and time framework for analyzing human anticipation of flash floods. *Journal of Hydrology,* 482, 14-24.

DE BRUIN, K., DELLINK, R., RUIJS, A., BOLWIDT, L., VAN BUUREN, A., GRAVELAND, J., DE GROOT, R., KUIKMAN, P., REINHARD, S. & ROETTER, R. 2009. Adapting to climate change in The Netherlands: an inventory of climate adaptation options and ranking of alternatives. *Climatic change,* 95, 23-45.

DE NEUFVILLE, R. 2002. Architecting/designing engineering systems using real options.

DE NEUFVILLE, R. & SCHOLTES, S. 2011. *Flexibility in engineering design*, MIT Press.

DE TONI, A. & TONCHIA, S. 1998. Manufacturing flexibility: a literature review. *International journal of production research,* 36, 1587-1617.

DERESSA, T. T., HASSAN, R. M. & RINGLER, C. 2011. Perception of and adaptation to climate change by farmers in the Nile basin of Ethiopia. *The Journal of Agricultural Science,* 149, 23-31.

DIFRANCESCO, K. N. & TULLOS, D. D. 2014. Flexibility in water resources management: review of concepts and development of assessment measures for flood management systems. *JAWRA Journal of the American Water Resources Association,* 50, 1527-1539.

DIFRANCESCO, K. N. & TULLOS, D. D. 2015. Assessment of flood management systems' flexibility with application to the Sacramento River basin, California, USA. *International Journal of River Basin Management,* 13, 271-284.

DORSCH, C. S. 2015. On the sound financial valuation of flexibility in information systems. *Business & Information Systems Engineering,* 57, 115-127.

DUINEN, R. V., FILATOVA, T., GEURTS, P. & VEEN, A. V. D. 2015. Empirical analysis of farmers' drought risk perception: Objective factors, personal circumstances, and social influence. *Risk analysis,* 35, 741-755.

EBRAHIMIAN, H. & LIAGHAT, A. 2011. Field evaluation of various mathematical models for furrow and border irrigation systems. *Soil Water Res,* 6, 91-101.

EEA 1999. Environmental indicators: typology and overview. *Technical Report No 25.* EEA, Copenhagen: European Environmental Agency.

ELASHA, B. O., ELHASSAN, N. G., AHMED, H. & ZAKIELDIN, S. 2005. Sustainable livelihood approach for assessing community resilience to climate change: case studies from Sudan. *Assessments of impacts and adaptations to climate change (AIACC) working paper.*

ERKOSSA, T. H., F.; LEFORE, N. (EDS.). 2014. Flood-based Farming for Food Security and Adaption to Climate Change in Ethiopia: Potential and Challenges. *In:* TEKLU ERKOSSA, F. H. A. N. L., ed., 30-31 October 2013 2014 Adama, Ethiopia,. International Water Management Institute (IWMI), P. O. Box 2075, Colombo, Sri Lanka: International Water Management Institute, 187p.

EVANS, J. S. 1991. Strategic flexibility for high technology manoeuvres: a conceptual framework. *Journal of management studies,* 28, 69-89.

FADUL, E., BASHIR, E., BUSHARA, A. & HAILE, A. M. 2012. Sharing experience among water user associations in spate irrigated schemes. Medani-Sudan: Hydraulic Research Centre, Ministry of Water Resources and Electricity.

FADUL, E., FRAITURE, C. D. & MASIH, I. 2018. Risk Propagation in Spate Irrigation Systems: A Case Study from Sudan. *Irrigation and Drainage,* 67, 363-373.

FADUL, E., MASIH, I. & FRAITURE, C. D. 2019. Adaptation strategies to cope with low, high and untimely floods: Lessons from the Gash spate irrigation system, Sudan. *Agricultural Water Management,* 217, 212-225.

FAO 1988. Irrigation water management: irrigation methods. *Training manual.* Via delle Terme di Caracalla, 00100 Rome, Italy.: Food and Agriculture Organization of the United Nations.

FAO. 2016. *AQUASTAT website* [Online]. Food and Agriculture Organization of the United Nations. [Accessed 10022016].

FAO 2018. The State of Food Security and Nutrition in the World. *Building Climate Resilience for Food Security and Nutrition.* Rome: Food and Agriculture Organization of the United Nations.

FAO AQUASTAT 2010. FAO–AQUASTAT.FAO's information system on water and agriculture.

FINLEY, S. 2016. *Sustainable Water Management in Smallholder Farming: Theory and Practice,* CABI.

FLOODSITE 2009. FLOODsite; Integrated Flood Risk Analysis and Management Methodologies. *CD-Rom. FLOODsite Consortium.* Delft, The Netherlands.: Deltares.

FOSU-MENSAH, B. Y., VLEK, P. L. & MACCARTHY, D. S. 2012. Farmers' perception and adaptation to climate change: a case study of Sekyedumase district in Ghana. *Environment, Development and Sustainability,* 14, 495-505.

FREDERICK, K. D., MAJOR, D. C. & STAKHIV, E. Z. 1997. Water resources planning principles and evaluation criteria for climate change: summary and conclusions. *Climate Change and Water Resources Planning Criteria.* Springer.

FUNK, C., DETTINGER, M. D., MICHAELSEN, J. C., VERDIN, J. P., BROWN, M. E., BARLOW, M. & HOELL, A. 2008. Warming of the Indian Ocean threatens eastern and southern African food security but could be mitigated by agricultural development. *Proceedings of the national academy of sciences,* 105, 11081-11086.

GAYDON, D., MEINKE, H. & RODRIGUEZ, D. 2012. The best farm-level irrigation strategy changes seasonally with fluctuating water availability. *Agricultural water management,* 103, 33-42.

GELETA, Y. 2014. Overview of challenges and opportunities of spate irrigation development in Oromia region. *Flood-based Farming for Food Security and Adaption to Climate Change in Ethiopia: Potential and Challenges,* 117.

GERWIN, D. 1993. Manufacturing flexibility: a strategic perspective. *Management science,* 39, 395-410.

GHEBREMARIAM, B. H. & STEENBERGEN, F. 2007. Agricultural water management in ephemeral rivers: community management in spate irrigation in Eritrea. *African Water Journal,* 1, 49-65.

GIEWS 2018. Crop Prospects and Food Situation. *GIEWS - Global Information and Early Warning System on Food and Agriculture.* Viale delle Terme di Caracalla, 00153 Rome - Italy: FAO.

GILLIES, M., SMITH, R., WILLIAMSON, B. & SHANAHAN, M. Improving performance of bay irrigation through higher flow rates. Australian Irrigation Conference and Exibition 2010: Proceedings, 8-11 June 2010 2010 Sydney, Australia. Irrigation Australia Ltd., 12pp.

GILLIES, M. H. & SMITH, R. J. 2015. SISCO: surface irrigation simulation, calibration and optimisation. *Irrigation Science,* 33, 339-355.

GIUPPONI, C. 2007. Decision Support Systems for implementing the European Water Framework Directive: The MULINO approach. *Environmental Modelling & Software,* 22, 248-258.

GIUPPONI, C., MYSIAK, J., FASSIO, A. & COGAN, V. 2004. MULINO-DSS: a computer tool for sustainable use of water resources at the catchment scale. *Mathematics and Computers in Simulation,* 64, 13-24.

GOLDEN, W. & POWELL, P. 1999. Exploring inter-organisational systems and flexibility in Ireland: a case of two value chains. *International journal of agile management systems,* 1, 169-176.

GONZÁLEZ-CEBOLLADA, C., MORET-FERNÁNDEZ, D., CERVERA-BIELSA, L. & MARTÍNEZ-CHUECA, V. 2011. Optimization of soil surface to save water in surface irrigation. 15pp.

GONZÁLEZ, C., CERVERA, L. & MORET-FERNÁNDEZ, D. 2011. Basin irrigation design with longitudinal slope. *Agricultural Water Management,* 98, 1516-1522.

GUNDERSEN, J. 2016. *Internally displaced persons in Sudan: a study of livelihood and coping strategies under protracted displacement.* MSc., Norwegian University of Life Sciences.

GUPTA, Y. P. & GOYAL, S. 1989. Flexibility of manufacturing systems: concepts and measurements. *European journal of operational research,* 43, 119-135.

HAILE A., S. B., DEPEWEG H. 2003. Water sharing and conflicts in the Wadi Laba spate irrigation system, Eritrea. *ICID conference* Montpellier.

HAILE, A. M. 2007. *A Tradition in Transition, Water Management Reforms and Indigenous Spate Irrigation Systems in Eritrea: PhD, UNESCO-IHE Institute for Water Education, Delft, The Netherlands*, CRC Press.

HAILE, A. M. 2010. Meeting Climate Change and Food Security Challenges in Fragile States. *Spate Irrigation Network Publications, Practical Note 1* [Online].

HAILE, A. M., DEPEWEG, H. & SCHULTZ, B. 2005a. Hydraulic performance evaluation of the Wadi Laba spate irrigation system in Eritrea. *Irrigation and drainage,* 54, 389-406.

HAILE, A. M., SCHULTZ, B. & DEPEWEG, H. 2003. Water Sharing and Conflicts in the Wadi Laba Spate Irrigation System, Eritrea. Spate irrigation network.

HAILE, A. M., SCHULTZ, B. & DEPEWEG, H. 2005b. Where indigenous water management practices overcome failures of structures: the Wadi Laba spate irrigation system in Eritrea. *Irrigation and drainage,* 54, 1-14.

HAILE, A. M., SCHULTZ, B. & DEPEWEG, H. 2006. Salinity impact assessment on crop yield for Wadi Laba spate irrigation system in Eritrea. *Agricultural Water Management,* 85, 27-37.

HAILE, A. M., SCHULTZ, B., DEPEWEG, H. & LAAT, P. D. 2008. Modelling soil moisture and assessing its impacts on water sharing and crop yield for the Wadi Laba spate irrigation system, Eritrea. *Irrigation and drainage,* 57, 41-56.

HAILE, A. M., STEENBERGEN, F. V. & SCHULTZ, B. 2005c. Water rights and rules, and management in spate irrigation systems. *African Water Laws: Plural Legislative Frameworks for Rural Water Management in Africa.* Johannesburg, South Africa.

HAILE, A. M., STEENBERGEN, F. V. & SCHULTZ, B. 2007. Water Rights and Rules, and Management in Spate Irrigation Systems in Eritrea, Yemen and Pakistan. *Community-based water law and water resource management reform in developing countries,* 114.

HAILE, A. M., STEENBERGEN, F. V. & SCHULTZ, B. 2011. Modernization of spate irrigated agriculture: A new approach. *Irrigation and drainage,* 60, 163-173.

HCENR 2003. Sudan's First National Communications under the United Nations Framework Convention on Climate Change *In:* MINISTRY OF ENVIRONMENT & PHYSICAL DEVELOPMENT & RESOURCES, H. C. F. E. A. N. (eds.).

HIBEN, M. G. & TESFA-ALEM, G. 2014a. Spate irrigation in Tigray: the challenges and suggested ways to overcome them. *Flood-based Farming for Food Security and Adaption to Climate Change in Ethiopia: Potential and Challenges,* 137.

HIBEN, M. G. & TESFA-ALEM, G. 2014b. Spate irrigation in Tigray: the challenges and suggested ways to overcome them. *Flood-based Farming for food security and adaption to climate change in Ethiopia: potential and challenges, International Water Management Institute (IWMI). Colombo, Sri Lanka,* 137-148.

HORRILLO-CARABALLO, J. M., REEVE, D. E., SIMMONDS, D., PAN, S., FOX, A., THOMPSON, R., HOGGART, S., KWAN, S. S. & GREAVES, D. 2013. Application of a source-pathway-receptor-consequence (SPRC) methodology to the Teign Estuary, UK. *Journal of Coastal Research,* 2, 1939-1944.

HRC 2016. On farm water management in Gash Agricultural Scheme. *From Africa to Asia and Back Again: Testing Adaptation in Flood-based Livelihood Systems* Wad Medani, Sudan: Hydraulic Research Centre.

HUANG, D., VAIRAVAMOORTHY, K. & TSEGAYE, S. Flexible design of urban water distribution networks. World Environmental and Water Resources Congress 2010: Challenges of Change, 2010. 4225-4236.

IFAD 2003. Gash Sustainable Livelihoods Regeneration Project: Appraisal report.

IGLESIAS, A. & GARROTE, L. 2015. Adaptation strategies for agricultural water management under climate change in Europe. *Agricultural Water Management,* 155, 113-124.

IPCC 2007. IPCC, 2007: climate change 2007: impacts, adaptation and vulnerability. Contribution of working group II to the fourth assessment report of the intergovernmental panel on climate change. Cambridge University Press, Cambridge.

IRMAK, S., ODHIAMBO, L. O., KRANZ, W. L. & EISENHAUER, D. E. 2011. Irrigation efficiency and uniformity, and crop water use efficiency.

JONKMAN, S. N. 2005. Global perspectives on loss of human life caused by floods. *Natural hazards,* 34, 151-175.

KAMRAN, M. A. & SHIVAKOTI, G. P. 2013. Design principles in tribal and settled areas spate irrigation management institutions in P unjab, P akistan. *Asia Pacific Viewpoint,* 54, 206-217.

KHALID, A. H. A. 2009. *Influence of Hydraulic Behaviour of Gash Delta Soil On Water Management.* Master of Science, Gezira University.

KHAN, S. R., NAWAZ, K., STEENBERGEN, F. V., NIZAMI, A. & AHMAD, S. 2014. The Dry Side of the Indus. Exploring Spate Irrigation in Pakistan. Spate Irrigation Network.

KOECH, R., GILLIES, M. & SMITH, R. Simulation modelling in surface irrigation systems. Proceedings of the 2010 Southern Region Engineering Conference (SREC 2010), 2010. Engineers Australia, 156-163.

KOMAKECH, H. C., MUL, M. L., VAN DER ZAAG, P. & RWEHUMBIZA, F. B. 2011. Water allocation and management in an emerging spate irrigation system in Makanya catchment, Tanzania. *Agricultural Water Management,* 98, 1719-1726.

KOPPEN, B. C., GIORDANO, M. & BUTTERWORTH, J. 2007. *Community-based water law and water resource management reform in developing countries,* London, UK, CABI.

KOWSAR, S. 2011a. Flood water spreading and spate irrigation in Iran. *Spate Irrigation Network,* 20.

KOWSAR, S. 2011b. Flood water spreading and spate irrigation in Iran. *Overview paper, practical note No 7.* Spate Irrigation Network.

LAWRENCE, P. & VAN STEENBERGEN, F. 2005. Improving community spate irrigation. *Report OD154 release 1.0.* London-UK HR Wallingford/DFID/META META.

LEMPERT, R. J. 2003. *Shaping the next one hundred years: new methods for quantitative, long-term policy analysis,* Rand Corporation.

LIBSEKAL, H., GEBRE-EGZIABHER, T., MEZGEBU, A. & YAZEW, E. 2015. Developments in the design and construction of modern spate systems in Tigray, Ethiopia. *Civil Environmental Research,* 7, 108-113.

LINQUITI, P. & VONORTAS, N. 2012. The value of flexibility in adapting to climate change: a real options analysis of investments in coastal defense. *Climate Change Economics,* 3, 1250008.

LUCIO, P. S., MOLION, L. C. B., DE AVILA VALAD, C., CONDE, F. C., RAMOS, A. M. & DE MELO, M. L. D. 2012. Dynamical outlines of the rainfall variability and the ITCZ role over the West Sahel.

MADDISON, D. 2007. *The perception of and adaptation to climate change in Africa,* World Bank Publications.

MAJOR, D. C. & FREDERICK, K. D. 1997. Water resources planning and climate change assessment methods. *Climate Change and Water Resources Planning Criteria.* Springer.

MARCHI, L., BORGA, M., PRECISO, E. & GAUME, E. 2010. Characterisation of selected extreme flash floods in Europe and implications for flood risk management. *Journal of Hydrology,* 394, 118-133.

MARTÍNEZ-ALVAREZ, V., GARCÍA-BASTIDA, P., MARTIN-GORRIZ, B. & SOTO-GARCÍA, M. 2014. Adaptive strategies of on-farm water management under water supply constraints in south-eastern Spain. *Agricultural water management,* 136, 59-67.

MCHARO, A. C. 2013. *Perception of farmers on effectiveness of agricultural extension agents in knowledge transfer to maize growers in Kilindi district.* Sokoine University Of Agriculture.

MERTZ, O., MBOW, C., REENBERG, A. & DIOUF, A. 2009. Farmers' perceptions of climate change and agricultural adaptation strategies in rural Sahel. *Environmental management,* 43, 804-816.

MIKAELIAN, T., NIGHTINGALE, D. J., RHODES, D. H. & HASTINGS, D. E. 2011. Real options in enterprise architecture: a holistic mapping of mechanisms and types for uncertainty management. *IEEE Transactions on Engineering Management,* 58, 457-470.

MOHAMED, N. A. H., BANNARI, A., FADUL, H. M. & ZAKIELDEEN, S. 2016. Ecological Zones Degradation Analysis in Central Sudan during a Half Century Using Remote Sensing and GIS. *Advances in Remote Sensing,* 5, 355.

MU'ALLEM, A. S. 1987. Crop production under spate irrigation in coastal areas of PDRY. *Proceedings of the Subregional Expert Consultation on Wadi Development for Agriculture in the Natural Yemen 6-10 December, 1987 Aden, PDR Yemen. Rome: FAO/UNDP.*

MULI, C., GERBER, N., SAKKETA, T. & MIRZABAEV, A. 2018. Ecosystem tipping points due to variable water availability and cascading effects on food security in Sub-Saharan Africa. *In:* CHRISTIAN BORGEMEISTER, J. V. B., MANFRED DENICH, TILL STELLMACHER AND EVA YOUKHANA (ed.). Bonn, Germany: Centre for Development Research, University of Bonn.

MYERS, R. 1980. The root system of a grain sorghum crop. *Field Crops Research,* 3, 53-64.

MYERS, S. C. 1977. Determinants of corporate borrowing. *Journal of financial economics,* 5, 147-175.

NARAYAN, S., HANSON, S., NICHOLLS, R. & CLARKE, D. 2011. Use of the source-pathway-receptor-consequence model in coastal flood risk assessment. *European Geosciences Union General Assembly,* 13.

NARAYAN, S., HANSON, S., NICHOLLS, R., CLARKE, D., WILLEMS, P., NTEGEKA, V. & MONBALIU, J. 2012. A holistic model for coastal flooding using system diagrams and the Source–Pathway–Receptor (SPR) concept. *Natural Hazards and Earth System Science,* 12, 1431-1439.

NARAYAN, S., NICHOLLS, R. J., CLARKE, D., HANSON, S., REEVE, D., HORRILLO-CARABALLO, J., LE COZANNET, G., HISSEL, F., KOWALSKA, B. & PARDA, R. 2014. The SPR systems model as a conceptual foundation for rapid integrated risk appraisals: Lessons from Europe. *Coastal Engineering,* 87, 15-31.

NARAYAN, S., SIMMONDS, D., NICHOLLS, R. J. & CLARKE, D. 2015. A Quasi-2D Bayesian network model for assessments of coastal inundation pathways and probabilities. *Journal of Flood Risk Management*, 1-28.

NGIRAZIE, L. A., BUSHARA, A. I. & KNOX, J. W. 2015. Assessing the performance of water user associations in the Gash Irrigation Project, Sudan. *Water International,* 40, 635-646.

NICHOLAS, K. A. & DURHAM, W. H. 2012. Farm-scale adaptation and vulnerability to environmental stresses: Insights from winegrowing in Northern California. *Global Environmental Change,* 22, 483-494.

NIE, W., FEI, L. & MA, X. 2014. Impact of infiltration parameters and Manning roughness on the advance trajectory and irrigation performance for closed-end furrows. *Spanish Journal of Agricultural Research,* 12, 1180-1191.

NILES, M. T. & MUELLER, N. D. 2016. Farmer perceptions of climate change: Associations with observed temperature and precipitation trends, irrigation, and climate beliefs. *Global Environmental Change,* 39, 133-142.

OOSTERBAAN, R. Spate Irrigation: Water harvesting and agricultural land development options in the NWFR of Pakistan. Proceeding of International Policy Workshop "Water Management and Land Rehabilitation, NW Frontier Region, Pakistan", Islamabad, 2010. Citeseer.

ORTEGA-REIG, M., PALAU-SALVADOR, G., I SEMPERE, M. J. C., BENITEZ-BUELGA, J., BADIELLA, D. & TRAWICK, P. 2014. The integrated use of surface, ground and recycled waste water in adapting to drought in the traditional irrigation system of Valencia. *Agricultural water management,* 133, 55-64.

OSMAN-ELASHA, B., GOUTBI, N., SPANGER-SIEGFRIED, E., DOUGHERTY, B., HANAFI, A., ZAKIELDEEN, S., SANJAK, A., ATTI, H. A. & ELHASSAN, H. M. 2006. Adaptation strategies to increase human resilience against climate variability and change: Lessons from the arid regions of Sudan. *Assessments of Impacts and Adaptations to Climate Change (AIACC) Working Paper* [Online]. Available: http://www.start.org/Projects/AIACC_Project/working_papers/Working%20Papers/AIACC_WP42_Osman.pdf.

OUDRA, I. 2011. Spate irrigation in Morrocco. *Overview paper No 6.* Spate Irrigation Network.

PALMER, M. A., LETTENMAIER, D. P., POFF, N. L., POSTEL, S. L., RICHTER, B. & WARNER, R. 2009. Climate Change and River Ecosystems: Protection and Adaptation Options. *Environmental Management,* 44, 1053-1068.

PALMER, R. & O'KEEFE, S. 2007. Irrigation Futures of the Goulburn Broken Catchment. *In:* TEAM, U. A. P. L. A. I. F. P. (ed.) *Handbook of flexible technologies for irrigation infrastructure.* Tatura, Australia Department of Primary Industries, Future Farming Systems Research Division.

PERKS, M. T., RUSSELL, A. J. & LARGE, A. R. 2016. Advances in flash flood monitoring using unmanned aerial vehicles (UAVs). *Hydrology and Earth System Sciences,* 20, 4005-4015.

PLAYÁN, E., WALKER, W. & MERKLEY, G. 1994. Two-dimensional simulation of basin irrigation. I: Theory. *Journal of irrigation and drainage engineering,* 120, 837-856.

PRADHAN, N. S., FU, Y., ZHANG, L. & YANG, Y. 2017. Farmers' perception of effective drought policy implementation: A case study of 2009–2010 drought in Yunnan province, China. *Land Use Policy,* 67, 48-56.

QUIROGA, S. & IGLESIAS, A. 2009. A comparison of the climate risks of cereal, citrus, grapevine and olive production in Spain. *Agricultural Systems,* 101, 91-100.

REES, D. 1987. Tactical operation of rice irrigation systems during water shortages. *Irrigation and Water Allocation. International Association of Hydrological Sciences Press, Institute of Hydrology, Wallingford, Oxfordshire UK. IAHS Publication.*

ROZALIS, S., MORIN, E., YAIR, Y. & PRICE, C. 2010. Flash flood prediction using an uncalibrated hydrological model and radar rainfall data in a Mediterranean watershed under changing hydrological conditions. *Journal of Hydrology,* 394, 245-255.

SABBAGHIAN, R. J., ZARGHAMI, M., NEJADHASHEMI, A. P., SHARIFI, M. B., HERMAN, M. R. & DANESHVAR, F. 2016. Application of risk-based multiple criteria decision analysis for selection of the best agricultural scenario for effective watershed management. *Journal of Environmental Management,* 168, 260-272.

SABINE, C. 2014. The IPCC fifth assessment report. *Carbon.*

SAHER, F. N., NASLY, M., ASMAWATY, T., ABDUL, B. & YAHAYA, N. 2014. Harnessing floodwater of hill torrents for improved spate irrigation system using geo-informatics approach. *Research Journal of Recent Science,* 3, 14-22.

SALAHOU, M. K., JIAO, X. & LÜ, H. 2018. Border irrigation performance with distance-based cut-off. *Agricultural Water Management,* 201, 27-37.

SALAZAR, L., TOLISANO, J., CRANE, K., WHEELER, L., KUILE, M. T. & RADTKE, D. 1994. Irrigation Reference Manual. *In:* (ICE), P. C. I. C. E. (ed.). 1111 - 20th Street, NW, Washington, DC 20526, USA: Agro Engineering, Inc., 0210 Road 2 South, Alamosa, Colorado 81101.

SALEH, J. H., HASTINGS, D. E. & NEWMAN, D. J. 2003. Flexibility in system design and implications for aerospace systems. *Acta astronautica,* 53, 927-944.

SCHULZ, A. P., FRICKE, E. & IGENBERGS, E. Enabling Changes in Systems throughout the Entire Life-Cycle–Key to Success? INCOSE international symposium, 2000. Wiley Online Library, 565-573.

SHIVAKOTI, G. P. & THAPA, S. B. 2005. Farmers' perceptions of participation and institutional effectiveness in the management of mid-hill watersheds in Nepal. *Environment and Development Economics,* 10, 665-687.

SIDDIG, K. H. & BABIK, B. I. 2017. Agricultural Efficiency Gains and Trade Liberalization in Sudan. *African Journal of Agricultural and Resource Economics,* 1.

SIDDIG, K. H. & BABIKER, B. I. 2012. Agricultural efficiency gains and trade liberalization in Sudan. *African Journal of Agricultural and Resource Economics,* 7, 51.

SIN 2013a. Spate irrigation in Mynamar. *Overview paper No 9.* Spate irrigation Network.

SIN 2013b. Traditional spate irrigation system in Al-Hajjareen – Dawan – Hadramut. *In:* SALEH, S. A. A. (ed.) *Overview paper No 21* Yemen: Spate Irrigation Network.

SLACK, N. 1983. Flexibility as a manufacturing objective. *International Journal of Operations & Production Management,* 3, 4-13.

SMITH, R., GILLIES, M. H., SHANAHAN, M., CAMPBELL, B. & WILLIAMSON, B. Evaluating the performance of bay irrigation in the GMID. Irrigation Australia 2009: Irrigation Australia Irrigation and Drainage Conference: Proceedings, 2009. Irrigation Australia Ltd., 1-12.

SPENCE, A., POORTINGA, W., BUTLER, C. & PIDGEON, N. F. 2011. Perceptions of climate change and willingness to save energy related to flood experience. *Nature climate change,* 1, 46-49.

SPILLER, M., VREEBURG, J. H., LEUSBROCK, I. & ZEEMAN, G. 2015. Flexible design in water and wastewater engineering–definitions, literature and decision guide. *Journal of Environmental Management,* 149, 271-281.

STEENBERGEN, F., LAWRENCE, P., MEHARI, A., SALMAN, M. & FAURÉS, J. 2010. Guidelines on spate irrigation. FAO Irrigation and Drainage Paper 65. ISBN 978-92-5-106608-09, FAO, Rome, Italy.

STEENBERGEN, F. V. & HAILE, A. M. 2010. Command Area Improvement and Soil Moisture Conservation in Spate Irrigation. Pakistan: Spate irrigation Network.

STEENBERGEN , F. V., LAWRENCE, P., HAILE, A. M., SALMAN, M. & FAURES, J. 2011. Guidelines on spate irrigation. *FAO irrigation and drainage paper,* 65.

STEENBERGEN, F. V., MEHARI, A. H. & ANDERSON, I. M. 2011. SPATE IRRIGATION IN BLUE NILE COUNTRIES: STATUS AND POTENTIAL.

STEVENSON, M. & SPRING, M. 2007. Flexibility from a supply chain perspective: definition and review. *International journal of operations & production management,* 27, 685-713.

STRELKOFF, T. 1985. BRDRFLW: A mathematical model of border irrigation. *In:* U.S. DEPT. OF AGRICULTURE, A. R. S. (ed.). Washington, D.C.: ARS (USA).

STRELKOFF, T. 1990. *SRFR: A computer program for simulating flow in surface irrigation: Furrows-basins-borders,* USDA-ARS, Phoenix, Arizona-USA, Water Conservation Laboratory, Agricultural Research Service, US Department of Agriculture.

STRELKOFF, T., CLEMMENS, A. & SCHMIDT, B. 1998. SRFR, Version 3.31—A model for simulating surface irrigation in borders, basins and furrows. *In:* US DEPARTMENT OF AGRICULTURE AGRICULTURAL RESEARCH SERVICE, U. W. C. L. (ed.). Broadway, Phoenix, Arizona-USA.

TADESSE, K. B. & DINKA, M. O. 2018. Improving Traditional Spate Irrigation Systems: A Review. *Landscape Architecture-The Sense of Places, Models and Applications.* IntechOpen.

TAGHIZADEH, Z., VERDINEJAD, V., EBRAHIMIAN, H. & KHANMOHAMMADI, N. 2013. Field Evaluation and Analysis of Surface

Irrigation System with WinSRFR (Case study- Furrow Irrigation). *Journal of Water and Soil (Agricultural Science and Technology)* 26, 1450 To 1459.

TALJAARD, S., SLINGER, J. H. & VAN DER MERWE, J. H. 2011. Criteria for evaluating the design of implementation models for integrated coastal management. *Coastal management,* 39, 628-655.

TESFAI, M. & STROOSNIJDER, L. 2001. The Eritrean spate irrigation system. *Agricultural water management,* 48, 51-60.

TOUILI, N., BAZTAN, J., VANDERLINDEN, J.-P., KANE, I. O., DIAZ-SIMAL, P. & PIETRANTONI, L. 2014. Public perception of engineering-based coastal flooding and erosion risk mitigation options: Lessons from three European coastal settings. *Coastal Engineering,* 87, 205-209.

TRIANTIS, A. J. 2000. Real options and corporate risk management. *Journal of applied corporate finance,* 13, 64-73.

TRIANTIS, A. J. 2003. Real options. *Handbook of modern finance,* D1-D32.

TRIGEORGIS, L. 2005. Making use of real options simple: An overview and applications in flexible/modular decision making. *The Engineering Economist,* 50, 25-53.

TURRAL, H., BURKE, J. & FAURÈS, J.-M. 2011. Climate change, water and food security. Food and Agriculture Organization of the United Nations (FAO).

USAID 2016. Climate change risk profile- SUDAN. *ATLAS - Adaptation Thought Leadership and Assessments*

USDA, S. 1974. Soil Classification System. *Definition and Abbreviations for Soil Description. West Technical Service Center, Portland, Oregon, USA.*

VAN STEENBERGEN, F. 1997. Understanding the sociology of spate irrigation: cases from Balochistan. *Journal of arid environments,* 35, 349-365.

VAN STEENBERGEN, F., LAWRENCE, P., MEHARI, A., SALMAN, M. & FAURÉS, J. 2010. Guidelines on spate irrigation. FAO Irrigation and Drainage, Paper 65. Rome, Italy: Food and Agricultural Organization

VAN STEENBERGEN, F., MACANDERSON, I. & MEHARI, A. 2011. Spate irrigation in the Horn of Africa: Status and potential. *Spate Irrigation Network Overview paper,* 2.

VAN STEENBERGERN, F., BAMAGA, O. & AL-WESHALI, A. 2011. Groundwater security in Yemen: Who is accountable to whom. *Law Env't & Dev. J.,* 7, 164.

VOLBERDA, H. W. 1996. Toward the flexible form: How to remain vital in hypercompetitive environments. *Organization science,* 7, 359-374.

WALKER, W. 1998. SIRMOD, a surface irrigation modelling software. *In:* ENGINEERING, D. O. B. A. I. (ed.). UTAH State University, Logan, USA.

WALKER, W. R. 2003. SIRMOD III: Surface irrigation simulation, evaluation and design. *Guide and Technical Documentation. Department of Biological and Irrigation Engineering. Utah State University, Logan, UT, USA.*

WARD, M. J., FERRAND, Y. B., LAKER, L. F., FROEHLE, C. M., VOGUS, T. J., DITTUS, R. S., KRIPALANI, S. & PINES, J. M. 2015. The nature and necessity of operational flexibility in the emergency department. *Annals of emergency medicine,* 65, 156-161.

WOODWARD, M., GOULDBY, B., KAPELAN, Z., KHU, S. T. & TOWNEND, I. 2011. R eal O ptions in flood risk management decision making. *Journal of Flood Risk Management,* 4, 339-349.

ZAQHLOEL, L. 1987. Spate irrigation in Morocco. *Spate Irrigation, Proceedings of the the Sub–regional Expert Consultation on Wadi Development for Agriculture in the Natural Yemen. Aden, December 1987.*

ZARGHAMI, M. 2011. Effective watershed management; case study of Urmia Lake, Iran. *Lake and Reservoir Management,* 27, 87-94.

ZENEBE, T., MOHAMED, Y. & HAILE, A. M. 2015a. Mitigation of Sedimentation at the Diverstion Intake of Fota Spate Irrigation: Case Study of the Gash Spate Irrigation Scheme, Sudan. *Irrig. Drain. Syst. Eng.,* 4, 1000138.

ZENEBE, T. F., HAILE, A. M. & MOHAMED, Y. 2015b. Mitigation of Sedimentation at the Diverstion Intake of Fota Spate Irrigation: Case Study of the Gash Spate Irrigation Scheme, Sudan. *Irrigation & Drainage Systems Engineering,* 04.

ZERIHUN, D., SANCHEZ, C., FARRELL-POE, K. & YITAYEW, M. 2005. Analysis and design of border irrigation systems. *Transactions of the ASAE,* 48, 1751-1764.

ZHANG, S., XU, D., LI, Y. & CAI, L. 2006. An optimized inverse model used to estimate Kostiakov infiltration parameters and Manning's roughness coefficient based on SGA and SRFR model:(I) establishment. *Journal of Hydraulic Engineering,* 11, 002.

ZIMMERER, K. S. 1995. The origins of Andean irrigation. *Nature,* 378, 481.

ZIMMERER, K. S. 2011. The landscape technology of spate irrigation amid development changes: Assembling the links to resources, livelihoods, and agrobiodiversity-food in the Bolivian Andes. *Global Environmental Change,* 21, 917-934.

LIST OF ACRONYMS

ha	hectare
m	Meter
hr(s)	Hour(s)
O&M	Operation and maintenance
CV	Coefficient of variation
SPRC	Source-Pathway-Receptors-Consequence model
DPSIR	Driving force-Pressure-State-Impact-Response
mDSS4	MULINO Decision Support System
AM	Analysis matrix
EM	Evaluation matrix
MCDM	Multi-criteria decision-making
LR	Low rank
HR	High rank
HA	Highly adopted
LA	Low adopted
T_{co}	Cut-off time
AE	Application efficiency
DU	Distribution uniformity
AD	Adequacy of irrigation
Q	Inflow rate
l/s	Liter per second
D_{req}	the required irrigation depth
U/S	Upstream
D/S	Downstream
M/S	Midstream
Misga	The smallest irrigation unit
CWR	Crop water requirement
DP	Deep percolation
RO	Run-off
WUAs	Water user associations
WM	Water managers
RMSE	Root mean square error
SF	Farmers 'measure to cope
SWUA	Water user associations' measure to cope
SM	Water managers' measure to cope
GW	Groundwater

ABOUT THE AUTHOR

Eiman Fadul graduated from Sudan University of Science and Technology with a Bachelor of Technology in Civil Engineering, specialization Hydraulics in 1997. After graduation, she joined Hydraulic Research Centre in The Ministry of Irrigation & Water Resources as a research Engineer. Her main responsibilities conducting applied research related to design of irrigation schemes, sediment monitoring program for the Blue Nile, Nile rivers and irrigation canals, calibration studies of hydraulic structures.in addition to consultancy work for water projects. In 2000, she obtained her MSc.in Engineering Hydrology from National University of Ireland, Galway, Ireland. Her thesis was titled "Model Evaluation criteria". After MSc. She continued working at the Hydraulic Research Centre with more responsibilities in leading research projects and capacity building programs. In 2008 she was the winner from Sudan for the UNESCO Keizo Obuchi fund to conduct research on *Water Supply for Low Income Rural Communities'*. In 2009, she was a guest lecturer at Loughborough University in UK. Between 2010-2013 she was a country team leader for the Spate Irrigation for Rural Economic Growth and Poverty Alleviation Project funded by IFAD and UNESCO-IHE with the objectives to develop capacity building programmes based on solutions-oriented action research and documented practical experiences that in an evidence based manner contribute to rural poverty alleviation and accelerated growth in marginal areas, preparing technical reports for the project and disseminate project outputs. In May 2013, she started working on her PhD research at the Department of Water Science and Engineering at UNESCO-IHE. Her research topic was risk and coping strategies in spate irrigation system, cases study the Gash Agricultural Scheme in eastern Sudan. Her research was risks and coping strategies to uncertain water supply in spate irrigation.

JOURNALS PUBLICATIONS

Fadul, E., Masih, I. & De. Fraiture, C, 2019. Adaptation strategies to cope with low, high and untimely floods: Lessons from the Gash spate irrigation system, Sudan. Journal of Agricultural Water Management 217, 212-225.

Fadul, E., De. Fraiture, C., & Masih, I. 2019. Irrigation performance and under alternative field design in a spate irrigation system with large field dimensions. Paper submitted to Journal of Agricultural water management.

Fadul, E., Masih, I. & De. Fraiture, C, 2019. Flexibility to cope with risks in spate irrigation systems. To be submitted to Journal of Irrigation and Drainage.

Fadul, E., De. Fraiture, C., & Masih, I., 2018. 'Risk propagation in Spate irrigation systems: A case study from Sudan', Irrigation and Drainage.

Fadul, E., Bob Reed, 'Domestic Water Supply Options in Gezira Scheme', April 2010, Waterlines, Vol 29, No 2.

CONFERENCE PROCEEDINGS

Fadul, E., Avellino, J., & Masih, I., Alternative Strategy for Field Design in Spate Irrigated Agriculture: Gash Agriculture Scheme, A paper presented at the 4th African Regional Conference on Irrigation and Drainage, April 2016, Cairo, Egypt.

Fadul E., Indigenous Risk management Practices in Spate Irrigated Agriculture, 7th graduate studies and scientific research Conference, 20-23 February 2015, Khartoum, Sudan.

Fadul, E., Risk management for a productive and profitable spate irrigated agriculture, PhD symposium, UNESCO-IHE Institute for Water Education, Delft, Netherlands, 29-30 September 2014.

Fadul, E., Multidisciplinary Approach to control Schistosomiasis in Gezira Irrigation Scheme, accepted paper to be presented in 34th WEDC International Conference, Addis Ababa, Ethiopia, May 2009.

Gismalla, Y, Fadul, E., 'Design and remodelling of stable canals in Sudan, Scientific paper presented in International conference for Sediment Initiative, organized by UNESCO Chair of Water Resources, Khartoum Nov.2006.

Netherlands Research School for the
Socio-Economic and Natural Sciences of the Environment

D I P L O M A

For specialised PhD training

The Netherlands Research School for the
Socio-Economic and Natural Sciences of the Environment
(SENSE) declares that

Eiman Mohamed Fadul Bashir

born on 28 June 1973 in Wad Medani, Sudan

has successfully fulfilled all requirements of the
Educational Programme of SENSE.

Delft, 8 January 2020

The Chairman of the SENSE board

Prof. dr. Martin Wassen

the SENSE Director of Education

Dr. Ad van Dommelen

The SENSE Research School has been accredited by the Royal Netherlands Academy of Arts and Sciences (KNAW)

K O N I N K L I J K E N E D E R L A N D S E
A K A D E M I E V A N W E T E N S C H A P P E N

The SENSE Research School declares that Eiman Mohamed Fadul Bashir has successfully fulfilled all requirements of the Educational PhD Programme of SENSE with a work load of 51.9 EC, including the following activities:

SENSE PhD Courses

- Environmental research in context (2013)
- SENSE writing week (2015)
- Research in context activity: 'Initiating and organizing expert and stakeholder workshops on: Sharing experiences among Water User association in spate irrigated schemes (15-18 December 2012, Kassala Sudan)'

Other IHE Delft PhD and Advanced MSc Courses

- Where there is little data: how to estimate design variables in poorly gauged basins, (2014)
- Data acquision, pre-processing & modelling using PC Raster and Python framework (2015)
- Data acquisition, pre-processing & modelling using HEC-RAS, IHE Delft (2016)

External training at a foreign research institute

- Workshop "Implementation of drought early warning systems and developing the institutional framework for effective response in Africa", Waternet, South Africa (2013)

Management and Didactic Skills Training

- Supervising two MSc students with thesis (2012)
- Teaching in the MSc courses 'hydraulic and irrigation water management' (2012) and 'Spate irrigation and water management under drought and water scarcity' (2012)
- Organising stakeholder workshop on "Linking research to policy, capacity building and practice", Ministry of Water resources – Hydraulic Research Institute, Sudan (2012)
- Organizing annual meeting of the research project "Spate irrigation for rural economic growth and poverty alleviation", Ministry of Water Resources, Sudan (2012)

Selection of Oral Presentations

- *Alternative strategy for field design in spate irrigated agriculture case study : Gash agriculture scheme.* 4th African Regional Conference on Irrigation and Drainage (ARCID), 26-28 April 2016, Aswan, Egypt

SENSE Coordinator PhD Education

Dr. Peter Vermeulen